LOSING OUR COOL

ALSO BY STAN COX

Sick Planet: Corporate Food and Medicine

LOSING OUR COOL

Uncomfortable Truths About Our
Air-Conditioned World
(and Finding New Ways
to Get Through the Summer)

Stan Cox

THE NEW PRESS

NEW YORK
LONDON

Requests for permission to reproduce selections from this book should be mailed to:
Permissions Department, The New Press, 38 Greene Street, New York, NY 10013.

Published in the United States by The New Press, New York, 2010
Distributed by Perseus Distribution

CIP data available
ISBN 978-1-59558-489-2 (hc)

The New Press was established in 1990 as a not-for-profit alternative to the large,
commercial publishing houses currently dominating the book publishing industry.
The New Press operates in the public interest rather than for private gain, and is
committed to publishing, in innovative ways, works of educational, cultural, and
community value that are often deemed insufficiently profitable.

www.thenewpress.com

Composition by dix! Digital Prepress
This book was set in Minion

Printed in the United States of America

10   9   8   7   6   5   4   3   2   1

To Tom and Brenda Cox

CONTENTS

PREFACE

In a brief August 27, 2006, item, the *Topeka Capitol-Journal* announced, "The latest beneficiaries of global warming are the residents of Emporia [Kansas] and Lyon County. Their county has been selected as the site for the newest electric power plant in the Westar system." While acknowledging that many Kansans doubt the human role in global warming, the paper noted that unusually high summer temperatures, whatever their origin, had prompted area residents to burn through record amounts of energy to keep cool. Because the new plant would boost the local government's property tax revenues and create some jobs, the article could conclude on a high note: "Congratulations to Lyon County. That's what economic development is all about."

The power plant reached its full 600-megawatt capacity in the recession-plagued summer of 2009. Typically, an efficient natural-gas-fired power plant of that size running at full capacity emits approximately 250 tons of carbon dioxide each hour. By making its contribution to greenhouse gases, the plant will be helping to ensure continued high air-conditioning demand for the energy it generates, whatever the economic forecast.

In the pages that follow, I will not be arguing that air-conditioning has created the world of economic and ecological peril in which we have come to live. I do maintain that with energy at the root of the biggest crises we face, air-conditioning must be dealt with as a subject of debate, not as a fait accompli. To wrestle with the question of air-conditioning is to confront the staggering task we face in keeping the world habitable for humans.

University of South Florida history professor Raymond Arsenault told me at the end of 2008 that he'd been "almost embarrassed" a quarter-century earlier when his groundbreaking paper "The End of the Long, Hot Summer: The Air Conditioner and Southern Culture" was published. He told me that the value of investigating the broad implications of air-conditioning "seemed so obvious, but no one had taken the time to do it: it was almost like writing about the air." But today, his remains the most requested article ever published by the *Journal of Southern History*. The paper captivated me, not least because I grew up in Georgia during the years when air-conditioning was starting to spread quickly across the state. I recall the nights when my father positioned a sheet of plywood at the top of the stairs in our small apartment in an attempt to capture the precious air emanating from a window air conditioner and hold it upstairs, in the bedrooms. When I was twelve years old, we moved into a house equipped with central air, and I thought I'd been admitted to Paradise early. But by the time I'd finished high school three years after the first Earth Day, that cold, still, dry summertime air became, for me, laden with the many troubling associations that are the subject of this book.

Some of the ills that follow in the wake of air-conditioning— resource waste, climate change, ozone depletion, and disorientation of the human mind and body—call for cures more complex than simply producing more energy-efficient devices or more atmosphere-friendly refrigerants. Air-conditioning has also been an important tool in creating a society shot through with unsustainable trends: settlements of large human populations in fragile environments; an imbalance between indoor and outdoor life; buildings designed for dependence on high energy input; suburbanization, "mansionization," and the oversized car and commuter cultures; recklessly accelerated production and consumption; enhanced military power; and even the political shocks that have hit this country in recent decades. None of those trends will be reversed overnight.

I stopped living with air-conditioning when I moved from the South to the Midwest on my twenty-first birthday. I have spent most of my life since then in places with intensely hot summers—including central Kansas and southern India—and have not missed refrigerated air. Please don't misread me: I am not an ascetic, a Stoic, a Luddite, a miser, an "econag," or a person of unmeltable moral fiber. I've lived this way because I prefer it; for an explanation, I can do no better than quote the anonymous apartment resident who once told comfort researchers, "We don't use the air conditioner because it makes it too hot outside." It helps that the members of my family are not fond of air-conditioning; they have been active participants in this thermal experiment.

The story of air-conditioning does not feature a prominent corporate or political evildoer. Undoing the damage is not a simple matter of cracking down on wicked individuals or companies. Air-conditioning manufacturers and installers, construction companies, developers, automakers, electric utilities, and real-estate brokers are simply playing by the rules and filling the roles in the economy that are expected of them. It is those rules and roles that must be changed.

With air-conditioning now considered as natural as the air itself, I wrote this book to reopen the debate over whether our indoor environments should be refrigerated. After I wrote a short series of articles on air-conditioning for AlterNet.org in the summer of 2006, a friend suggested, "You're right about all that, but people just won't accept it. Why don't you go after an issue for which people are willing to change?" It was then that I decided to show that a convergence of evolving ideas with ecological and economic forces can bring radical change. When I told Ray Arsenault of my aim, he said, "You're sure to get an argument on that one. People expect things to be the way they want them to be, and they lose the ability to adapt. It's hard to think of a historical precedent, when people were willing to turn back the clock technologically."

As I began working on the book, I realized that the situation may be even more tangled than Arsenault suggested. To keep our demands on the planet within necessary bounds, we will be required to turn three kinds of clocks—technological, economic, and cultural—backward at times, forward at other times.

Two books have thoroughly examined the history of air-conditioning in the United States: Gail Cooper's *Air-Conditioning America: Engineers and the Controlled Environment, 1900–1960* and Marsha Ackermann's *Cool Comfort: America's Romance with Air-Conditioning.* Cooper provides a superb account of how technology, shelter design, and economics interacted in launching the age of air-conditioning. Ackermann's history covers the intersecting roles of culture, politics, and economics in creating the air-conditioned society we now live in. Their insights from the past provide a solid base from which to examine air-conditioning's future.

Air-conditioning's evolution from luxury to necessity in America is taken for granted by most of us, but a stubborn few have refused to accept it as inevitable. A big share of the criticism has come from outside our borders, often from countries that enjoy comfortable climates and may not fully appreciate the intensity of our summers. One of the most memorable moments in the debate over whether—rather than how—to air-condition came in mid-1992 with the publication of a special issue of the journal *Energy and Buildings*. The issue focused on the social and cultural aspects of cooling, and among the articles on alternative cooling methods and the logic of thermostat adjustment was Cambridge University professor Gwyn Prins's colorful, unconventional attack on air-conditioned "coolth" (the issue's editors observing with true academic understatement that "Prins' style of argument and evidence will be novel to many readers of this volume"). In it, he argued that "physical addiction to air-conditioned air is the most pervasive and least noticed epidemic in modern America." Comparing America to Africa, he described millions of "addicted" Americans as "by definition, cold and dry. They abhor the heat and the wet." Our craving for coolth (a real word indicating the approximate opposite of warmth), according to Prins, has led to a claim of an inalienable right:

> The stiff-necked, eighteenth-century Englishmen who composed the Declaration of Independence did not list "Life, Liberty and the Pursuit of Happiness" casually in that order. Category I is Life. Category II is Liberty and Category III is the Pursuit of Happiness: Ace, King, Queen. The problem about Coolth is that it makes a phony claim to be promoted from Category III to Category I.

Although I don't enjoy receiving lectures from the soft, green English countryside on heat tolerance and American revolutionary history, I do agree with Prins on his contention that air-conditioning does not deserve automatic promotion to the status of an essential national resource. But I refuse to concur with his characterization of our air-conditioned population as addiction victims. Nor will I join the "new moralists" of air-conditioning's early days who, as Ackermann discussed, were critical of the technology because it added to America's consumerist obsessions and undermined good character. And finally, I pledge not to sermonize about personal lifestyles. If, as I argue, air-conditioning is linked to a sprawling complex of environmental, social, and economic issues, the web of causes and consequences is not primarily personal but rather structural—in the political/economic sense as much as the sense of physical buildings. If, as Cooper and Ackermann showed, climate control was sold to an earlier generation without that generation even having asked for it, and if, as Prins argued and I have concluded, the net effect on the lives, happiness, and possibly even liberty of ordinary Americans has been negative, then rethinking the whole idea of air-conditioning could improve all of our lives.

Marsha Ackermann concluded her book with this sentence: "For better or worse, our world of tomorrow will be air-conditioned." The starting point for this book is that same statement—rephrased as a question.

ACKNOWLEDGMENTS

I am grateful to those editors who were willing to give me the benefit of the doubt when I told them that air-conditioning was a subject worth writing about: Tai Moses, Harris Rayl, Scott Bontz, and especially Jonathan Teller-Elsberg. I am grateful most of all to Sarah Fan for making the book possible and for the superb editing from which this book benefited so much. Raymond Dean was extremely generous with his time and deep knowledge of the technical aspects of air-conditioning. Those parts of the book could not have been written without him; if I got something wrong, the error is mine, not Ray's. Others who were exceptionally helpful in sharing their ideas and expertise were Raymond Arsenault, Gary Mormino, Christian Warren, Todd Bostwick, Dani Moore, Chris George, Bob Livingston, Darshan Bhatia, Girish Srinivasan, Girish Sant, Joshua Pearce, and David Wyon. I depended on Sheila Cox, as always, to do the initial critiques and editing. I am grateful to her, to Paul Cox, to my parents Tom and Brenda Cox, to Santosh Gulati, and to Priti Gulati Cox—the love of my life—for their patience and support.

1

"THERE'S NO POWER ON EARTH
THAT CAN STOP IT!"

*Air-conditioning and other technologies have made it so there's no place
on the planet where we couldn't live. We can condition the air; we can
make it right.*

—Jim Roberts, air-conditioning service manager,
Fort Myers, Florida, 2008

*Senator Ashurst used to tell on himself the story of his maiden speech in
the U.S. Senate. "Mr. President," he began, "the new baby state I represent has the greatest of potential. This state could become a paradise. We
need only two things: water and lots of good people." A gruff senior senator from New England interrupted, "If the senator will pardon me for
saying so, that's all they need in Hell!" We have lots of good people in Arizona. But after fifty years the search for dependable water supplies is still
the big story of our state.*

—Senator Morris Udall of Arizona, arguing for funding of the Central
Arizona Project to supply water from the Colorado River, 1963

You may have noticed, as I have, that when the heat of your surroundings reaches a near-intolerable intensity, it becomes almost audible. I
thought I started hearing the heat—as a growing, dull hum—one
windless July afternoon in Phoenix as I took a walk in the 114° sun.
The fact that I was strolling among the foundations of a village that
had last existed six hundred years ago (on land that now borders Phoenix Sky Harbor International Airport) caused further disorientation.
One obvious question still managed to penetrate the haze in my brain:
what sort of extraordinary beings had lived here and sustained a thriv-
ing civilization through a thousand such summers without modern
cooling technology?

From around 450 to 1450 A.D., the Hohokam people occupied this same central Arizona valley that now struggles to contain Phoenix. At their peak, they numbered between twenty-five and sixty thousand and set a record for population density in the desert Southwest that was broken only by European settlement centuries later. Throughout the Salt and Gila river valleys, they built a sophisticated system of canals with a total length of a thousand miles and cultivated a hundred thousand acres of crops. The village of Pueblo Grande—near the now-dry bed of the Salt River, where I wandered on that second-hottest afternoon of 2009—had been one of their many carefully planned housing developments.

The short, self-guided trail through Pueblo Grande is designed to take a half-hour under normal conditions, but I found myself speed-reading the plaques and finishing the trail in less than fifteen minutes. Quickly covering another few dozen steps, I ducked into the cool office of city archaeologist Todd Bostwick and asked the obvious question. Bostwick explained that survival without air-conditioning in this Valley of the Sun, as it's known, would not have required superhuman powers back in Hohokam days: "It's important to remember that the Salt River ran year-round then. They could go jump in the river or a canal to cool off whenever they wanted to. The canals were their roads, and [they] supported ribbons of green growth, which also provided cooling. They used wood, but they clearly appreciated the value of shade and would have left plenty of shade trees standing." Bostwick gestured through his window toward the village remains: "People didn't really live in those rooms. Those were chiefly for privacy. They did their work out in the shade of a ramada [an open, detached porch roofed by brush or branches]. They grew cotton and made cloth, so they probably would have employed the old trick of the wet sheet over the doorway for evaporative cooling. When it was too hot inside, they would have slept outdoors. And they were smart enough not to work in the heat of the day."

"Incidentally," Bostwick added, "I get that question a lot, the one about how the Hohokam got by without air-conditioning. But it always comes from adults. For some reason, kids never ask that. The notion of comfort is as cultural as it is personal."

Central Arizona's climate in Hohokam times was not very different from today's, says Bostwick—at most, it was just slightly cooler—but the landscape in this valley would certainly have been more tolerable without the vast expanse of heat-trapping concrete and asphalt that now covers a large portion of it. Could the valley ever have become home to millions of people without modern air-conditioning? Bostwick grinned. "Absolutely not. But here is my message: you don't need air-conditioning to build an incredibly successful culture in the desert. The Hohokam proved that. They proved it for more than a thousand years."

THE CAPITAL OF COOL

In 1940 the *Arizona Republic* crowned Phoenix "the air-conditioned capital of the world." That was more a prophecy than a description; mechanical evaporative cooling was still the predominant way of delivering comfort. But evaporative coolers (which update the Hohokam wet-sheet strategy by using simple fans to remove heat from the air, drawing it through water-soaked pads) were about to give way to increasingly cheap and efficient refrigerated air.

In that same article, the *Republic* declared, "Phoenicians do not move to new localities when they desire a climate change. They change the climate." That too, as we now know, was prophetic, but in a more dismal vein. Residents of Phoenix and the nearby cities and suburbs in Maricopa County are already living a future that awaits the rest of us. There, the decade-by-decade warming rate has been higher than in any of the world's other big cities—and Phoenix started off near the top of the charts. The urban "heat island" effect (through which the concrete, asphalt, and steel of cities absorb solar energy and hold it in the form of heat) has raised temperatures 7.6° Fahrenheit. That figure lies toward the high end of the range of temperature increases projected for the entire planet's greenhouse future. The city is also topped by a "carbon dioxide dome," with weekday concentrations of CO_2 "equivalent to what is being predicted for 100 years from now." Sixty-seven years after the *Republic* credited residents with changing the city's climate, the residents who followed were paying the price for decades of heroic growth. In 2007, Phoenix endured a record-breaking twenty-eight days with high temperatures exceeding 110°; compare that with an average of only 6.7 such days per summer in the 1950s. The deep nighttime temperature drop, that much-loved feature of the desert climate, had shrunk by one-third within the Phoenix city limits. Heat captured during the day was being trapped in concrete and asphalt and between buildings; as a result, Phoenix residents were having to keep air conditioners running longer into the evening and night hours. Matters reached an extreme on July 15–16, 2003, when the city "cooled" down only to 96° by early morning—fifteen degrees hotter than the long-term average low for the date—and hit 117° on the afternoon of the sixteenth.

Air-conditioning's environmental damage is not limited to emissions of greenhouse gases and ozone-depleting chemicals. It has also been used as a lever to open the Southwestern desert and other ecologically vulnerable parts of the country to reckless growth. Lavish deployment of indoor climate control may indeed make it possible for us to live anywhere on the planet, but is that wise? Across the southern tier of states, from the desert Southwest to the Everglades, air-conditioning has played an essential role in drawing millions of people to some of the country's most fragile environments as high-powered development steamrolls ecological barriers. In 1930, Phoenix had a population size half that of Peoria, Illinois. Now, fourteen times as many people live in Phoenix as in Peoria, even though the landscape and climate around Peoria have a much higher natural capacity to support a large population. (Incidentally, there is a Peoria, Arizona. The ninth-largest city in Arizona, it has been swallowed up by Phoenix sprawl.)

It took until 1920 for Maricopa County's population to regain and surpass its level of Hohokam times. Climate control arrived soon after, and by 2009, four million people were living in the county, most of them clustered in and around Phoenix. The county government expects six million by 2030. By the turn of the millennium, a powerful air-conditioning system, once a luxury, had come to serve as a life-support apparatus for Phoenicians, and not just because of the heat. As the *Los Angeles Times* reported in 1999:

Phoenix's air quality is well below national health standards, and its violations for particulates, ozone and carbon dioxide have been classified as "serious," a distinction shared with only one other city: Los Angeles. . . . Phoenix is teeming with those whose only job is to fight the dust battle. Pool cleaners, air-conditioning repairmen and house cleaners all report that their lives are made more miserable by dust, even as their livelihoods benefit. . . . Gaye Knight, the air quality advisor for the city of Phoenix, said her office receives 5,000 complaints a year about fugitive dust. . . . Valley fever, an illness whose symptoms can range from fatigue to fungus in bones or the lining of the brain, is a pathogen that lives quietly in the ground until soil is disturbed. The number of cases in Arizona has doubled in recent years, and the cost of hospitalizing valley fever patients has been put at more than $20 million. . . . A 1995 study estimated that 963 Arizonans a year die prematurely of respiratory ailments from inhaling particulates.

On the origin of those problems, the *Times* article quoted Howard Wilshire, a former senior scientist for the U.S. Geological Survey: "Undisturbed desert does not create large quantities of dust. A natural crust of algae and lichen forms in the desert and it stabilizes the soil. The problem is man. We are being exceedingly foolish with our abuse of the desert."

A square mile of desert is equipped to support small populations of appropriately adapted animal species, not 3,200 humans along with their homes, workplaces, and vehicles. Increased heat is also concentrating ozone and other components of smog in the city's air. Air conditioners are shielding residents from heat, dust, and smog but not from other health hazards. The increase in childhood and adult obesity has been accelerated, says recent research, because the ever-more-oppressive heat keeps kids confined to air-conditioned refuges. Life in a desert city involves many more, and often unexpected, perils.

By 2008, Phoenix politicians, academics, and business leaders were proudly pointing to an array of "green" initiatives that would prove that, despite all of those hazards, the central Arizona desert can tolerate a population of six million people or more. Plans included new reflective roofs and buildings, vegetation and urban forestry, better insulation, a "connected oasis" of street parks, and paving materials that conserve water by allowing it to percolate into the soil below. An attractive light-rail system is credited with putting a small dent in car traffic, but it consists of only a single line through Phoenix and neighboring Tempe and Mesa.

Environmental initiatives to date have yet to produce significant results, so some local residents are taking environmental matters into their own hands. Dani Moore and Chris George moved to the Phoenix area from New Jersey in midsummer 2006. As outdoor types, the newlyweds were happy to find themselves in a desert climate, and they decided that they could live year-round without air-conditioning. Throughout the summer of 2008, they largely stuck to their plan— except on a few occasions, as when they hosted out-of-town visitors or had to allow new flooring to set in. But in 2009, they resolved to leave the thermostat set to "Off" no matter what happened. And they did. I visited them in their small brick Tempe home during the hottest stretch of that summer, on a July day when the official Phoenix high was 113°, after a low the previous night of 91°. When I arrived at two o'clock, the temperature in their kitchen hovered around 100°. "Now we know we can make it through the summer without being tempted," said George. "The worst will soon be over." As we guzzled cold water, Moore, who was doing graduate work at Arizona State University in biology (specifically, in ant behavior) admitted that the original idea for their "air-conditioning strike" had been hers. "I love warmth anyway, and we found out that this kind of weather's not so bad." Had they thought about putting in a rooftop evaporative cooler? "We won't be living here long-term, so we are reluctant to invest in one," she said. "Besides, I'm pretty sure the homeowners' association here bans them on roofs." She said she'd found herself gazing longingly at a freestanding portable model in a store a few days before. "But I just don't think I want to have to deal with another appliance."

George was working at the *Arizona Republic* as a page designer, and he was documenting his experience through the summer in a blog on the paper's Web site. "I try not to be too preachy, but sometimes I can't help myself. And the purely practical economic argument about saving money doesn't always work." Despite tough times, he said, his blog received a lot of comments like "Well, *I'm* willing to pay for my comfort. Don't be a cheapskate." Other commenters were less polite, calling him crazy or, in one case, a "dysfunctional whack job." Whatever his own mental state, said George, his response to critics was "Do you really need to keep your house forty degrees below the outdoor temperature to be comfortable? Are you willing at least to raise your thermostat setting and try a ten- or twenty-degree difference? When you come inside from this kind of heat, that's enough to make you really feel cool." Some readers and friends responded by cutting back on their own resource use in various ways—turning up the thermostat, for example, or using a clothesline instead of an electric dryer—but he hadn't heard from anyone else who turned off the air-conditioning altogether.

Without air-conditioning at home (even though they both had it at work), Moore and George said their perception of temperature had become more flexible; as the weather warmed up through spring and summer, they felt comfortable at warmer and warmer temperatures. They agreed that "your blood thins when you come to Arizona." Going to and from work over the summer, George had found himself using his bicycle and the light-rail system most of the time rather than driving. When you wake up in a 90° house, he said, the breeze you create by riding a bike to work feels pretty good.

Dani Moore and Chris George are only two of many Valley of the Sun residents who are trying to live within tighter ecological limits. But have sixty-plus years of heedless expansion already exceeded the limits of human population that this desert valley can support? Not in the view of Arizona State University president Michael Crow, who asserted in 2008 that Phoenix has a golden opportunity to demonstrate to the world that sustainability is possible. That's because, he declared, there's going to be so much more growth with which to experiment: "When Phoenix is done growing, it will be bigger than Chicago. The next massive city of the United States isn't done yet."

PUSHING OUT THE DESERT

Former vice president Dan Quayle can be forgiven for having once announced, "I love California—I practically grew up in Phoenix." Air-conditioning has helped turn the city, along with its suburbs, into a place that's not easily distinguished from any other place in America. Before air-conditioning, Phoenix was an unassuming wintertime haven for tourists and retirees. During World War II, the city acquired several military bases, and local boosters took advantage of the area's comfortable winter weather, lack of troublesome humidity, and open spaces to reel in big business and industry. Writes Michael Logan in his 2006 book *Desert Cities: The Environmental History of Phoenix and Tucson*, "Critical to this effort was the assault on the summer heat. Tourists might be lured to the desert during the mild winters, but manufacturing firms required year-round residency." The city concentrated its efforts on bringing in "clean" industries, and succeeded early on in drawing General Electric, Sperry-Rand (now part of Unisys), and Motorola plants to the area. Neither the high-tech industries of that era nor those that came or expanded later—including communications, aviation, finance, semiconductors, health care, and, yes, air-conditioner manufacturing—would have considered locating in a Phoenix that could offer them and their employees only ceiling fans and swamp coolers to temper the heat and dust. In fact, housing construction, the area's biggest economic powerhouse, was the industry most hooked on air-conditioning.

In looking at the development of Phoenix in the 1960s and 1970s, Logan noticed a trend:

> On the one hand, resort owners and the tourist industry marketed desert landscapes and the dry, warm climate as positive community attributes. On the other hand, city dwellers and business leaders diligently sought to push the desert to the distant margins of the community's identity. . . . Shopping malls proliferated in the valley coincidentally with the expansion of the tourist resorts, but the commercial centers represent the community's effort to marginalize the desert. Shoppers would

drive into the massive parking lots in their air-conditioned cars and hustle across the bubbling asphalt to the chilled interior spaces of the new malls. Only in that brief sprint in and out of the stores would the shopper experience the desert climate. Inside the malls waterfalls gurgled and ice skaters twirled.

The selling of an alternative outdoors to desert-dwellers continues in today's postmall world. Holding down one corner of the mammoth new 250-acre Mesa Riverview shopping complex along the dry bed of the Salt River is a local branch of that supreme temple to the outdoor life, Bass Pro Shops. Striving to live up to its slogan "More outdoors for your money!" the Mesa outlet has packed a whole national park's worth of "outdoors" under its towering cathedral ceiling: populations of stuffed bears, deer, lions, water buffalo, and vultures; simulated rock formations and waterfalls; a flowing trout stream, complete with a stuffed, life-size fly fisherman; and an impressive aquarium filled with striped bass, channel catfish, bluegill, and other species. A chalkboard sign advises, "Trout feeding at 1:30." Customers can also inhale plenty of simulated mountain air for their money. With outdoor temperatures exceeding a dry 110° through much of the day, I'd found most area stores—even grocery stores—keeping their thermostats set above 80°. On the mezzanine level of Bass Pro Shops, it was 75°.

In the Phoenix valley, whatever the season, the first word to come to mind after "heat" is "water?"—always with a question mark attached. City officials declare that they have demand covered for the next hundred years, claiming that decades of practice in dealing with drought mean that Phoenix households already use less water than the national average. The resources, physical effort, and brainpower marshaled to supply the water demanded by the city's growth have been monumental, the crown jewel being the Central Arizona Project, a 336-mile-long system of aqueducts, tunnels, pumping plants, and pipelines bringing water from the Colorado River.

However extensive those efforts, the city always manages to test the limits of its water supply. Jay Golden, assistant professor at Arizona State University's School of Sustainability, has pointed out that the city's tight water situation could undermine efforts to deal with the heat because it can make tree planting impracticable. And xeriscaping—the use of desert plants, usually native ones, in place of lawns—saves much less water in practice than in theory. Some desert plants are adapted to staying alive with very little water, but when water's available, they have the ability to draw even more out of the soil than do water-intensive plants like mulberry trees. Meanwhile, for aesthetic reasons, residents tend to pack the vegetation into their yards at far higher densities than it is found in nearby natural landscapes. They also tend to water their xeric plants much more heavily than necessary. The high-curb-appeal result, what critics call "Disney Desert," is the runaway favorite style in middle-income front yards, according to a 2006 survey. Low-income residents prefer traditional green lawns, while the wealthy favor the "Oasis," also known as "California Light," in which "plants are selected for their brightly colored flowers and lush vegetation . . . the majority of plants are exotics and their density creates a tropical atmosphere." The Oasis requires daily irrigation. Researchers also found that the different types of vegetation are placed strategically: "In the visible front yard, desert landscaping (perceived by most as more socially correct) was the more frequent preference, while in the less visible backyard, the luscious, more water-consumptive oasis landscape was much more highly favored. Therefore, we propose that in the front yard, form follows fashion while in the backyard, form follows fantasy."

If the domestic landscapes of Phoenix boil down to form and fantasy, the homeowner may be able to enjoy them most when not experiencing them physically. In dashing out to check on their drip-irrigation systems and ducking back into their cool havens, present-day Phoenix homeowners might well agree with comments made by a Washington, D.C., couple interviewed in 1961 about their brand-new central-air system: "We enjoy gardening, but even more we enjoy being able to sit indoors comfortably and look out at our garden." Their sentiment was echoed forty-five years later in a confession that writer William Saletan inserted into an otherwise stinging critique of air-conditioning that ran in the *Washington Post*: "Seven years ago, when my wife and I moved into our house, we planted a garden and built a patio in the back yard. Now, overcome by heat and mosquitoes, we're thinking of replacing them with something a bit more climate-controlled. We still want to look at nature. We just don't want to feel it."

The state of Arizona gets almost 40 percent of its water from the Colorado River, which runs on its western border. By law, water from the river has to be shared with California, Nevada, and Mexico in a way that leaves at least some water for the natural ecosystems that depend on its flow. Projections show that, by 2050, Arizonans' demand for water from the already badly stressed Colorado will exceed what the river can provide by as much as 90 percent. On the positive side, urbanization has the potential to reduce water use. In traditionally agricultural Maricopa County, each acre converted from irrigated farmland to residential or commercial use brings a reduction in water consumption, because households use far less water per acre than does irrigated agriculture.

However, areas conquered by the urban march have not been limited to former farmland but also include desert and other nonirrigated lands. Therefore, a decrease in water consumption by crop irrigation in the county during the 1990s was partly canceled out by the 56 percent increase in domestic water use in rapidly growing urban and suburban areas. In addition, water consumption by single-family homes doubles in summertime, and the heat-island effect increases water consumption in the city center and suburbs by encouraging greater household water use and causing more loss through evaporation. And there are plenty of watery surfaces lying there ready to evaporate in a city where more than one of every four houses has a swimming pool.

SEIZING POWER

All of Arizona's neighboring states can boast of localities where temperatures reach withering highs. California has its Death Valley, Utah its Beaver Dam Wash, New Mexico its Carlsbad. But most people in those states tend to congregate, sensibly enough, in areas with more moderate climates. Nearly three-fourths of Nevada's population, on the other hand, is crowded near the state's hottest corner, in and around Las Vegas, while three out of five Arizonans sweat it out three hundred miles to the southeast in the Phoenix area. (To be fair, although it lies in one of its state's hottest spots, Phoenix does not hold the record for Arizona's highest recorded temperature. That distinction belongs to Lake Havasu City near the California state line, which has hit 128°, compared with a record of "only" 122° in Phoenix.)

The commonly used metric known as cooling degree-days (CDD) is the number of degrees that the daily average temperature exceeds 65°, added up through all days of the year. The National Climatic Data Center has weighted each state's CDD by its population distribution in order to compare the amount of hot weather experienced by the typical resident of each state. For California, Utah, and New Mexico, the weighted average is close to 1,000 CDD; for Nevada, it's about 2,600, and for Arizona, 3,403. Arizona runs a close second to Florida, which has the most thoroughly cooked population in the nation, with 3,650 CDD experienced by the average resident.

Heavy reliance on air-conditioning creates huge electricity demand in Arizona and Nevada. And the flow of electrical energy is growing faster in Arizona than in any other state. Consumption per person grew by 51 percent from 1981 to 2005 and now exceeds that of all the state's neighbors: 68 percent higher per capita consumption than New Mexico's, 70 percent more than Utah's, and 116 percent more than California's. Arizona has even surpassed Nevada and now uses 11 percent more electricity per capita than does its famously profligate neighbor to the north. Air-conditioning requires not only a lot of electricity, but vast volumes of gasoline and diesel as well. Running the air conditioners in cars and trucks throughout Arizona takes 216 million gallons of fuel annually.

The migration of millions of Americans over the past half-century from colder to warmer climates should, in theory, have reduced the total energy required for climate control. On average, home heating in the northern states consumes more units of energy, mostly from natural gas or fuel oil, than air-conditioning consumes, mostly as electricity, in the Sun Belt. As a population, Americans experienced an average of 4,822 heating degree-days (HDD, similarly computed, based on temperatures below 65°) in 1950; after more than a half-century of southward migration, that average heating requirement had shrunk to 4,260 HDD by 2007. Cooling needs rose over the same time from 1,217 to 1,424 CDD. The net result is that we now experience an average of 5,684 heating plus cooling degree-days annually, a 6 percent lower requirement for year-round climate control than the 6,039 degree-days we would be enduring if our population had simply grown to its present total without any southward migration. But, because Americans use vastly more air-conditioning than we did in 1950, we cancel out a big hunk of the energy savings that might have been achieved by the population's shift to the south. The net result is that we consume only 2.5 percent less total energy than we would be burning had those millions of Americans stayed up north heating millions more houses through long, cold winters. That said, there is a lot more to the nation's environmental health than the quantity of energy we use or how we get it. What we do with that energy is at least as crucial, and Phoenix is one of many examples that demonstrate how not to use it.

Arizona's famous sunshine is, potentially, a rich source of energy. Currently, about 3 percent of its residential electricity consumption is satisfied by solar generation; that's about double the national average. By one estimate, solar sources will become economically attractive enough by 2017 to start adding two to four hundred megawatts of new capacity to the state's power infrastructure every year. But that would not come close to satisfying summer demand, which has been growing by six hundred megawatts per year.

Seven percent of Arizona's electricity comes from hydroelectric sources. Power generated by dams is an especially useful contributor to the grid, responding almost instantly on hot afternoons and evenings when air-conditioning demand peaks. At such times, inlets can simply be opened wider, sending more water through the turbines. Before 1992, water flow in the Colorado River below the Glen Canyon Dam (whose turbines supply electricity to Arizona and other Southwestern states) could vary fourfold within a single summer day, reaching a peak as compressors in distant cities were pushed harder to cool homes, stores, and offices in the late afternoon. That could raise and lower the river level by more than twelve feet. At no other time in its history had the canyon experienced such large, repeated daily flushings. Damage to populations of plant and animal species downstream in Grand Canyon National Park was inevitable, and fluctuations in power output and water outflow from Glen Canyon Dam were moderated by law in 1992 to protect the park, but there are still significant ups and downs.

There is one true heavyweight power source in the Phoenix area. Head west into the desert, follow the biggest strands of power lines upstream, and you'll find the 3,875-megawatt Palo Verde nuclear generating plant. Its three white domes rise from the tan landscape and loom low against a sky that, thanks to dust, tends to be only a slightly paler shade of tan. Just as impressive as the domes covering the three reactor cores are a half-dozen enormous water-cooling towers. The plant, built to accommodate expected growth in the region, commenced operation in 1988, and its output helps ensure that Phoenix's air conditioners will keep running as summer temperatures continue to climb. In making that cooling possible, the plant generates its own huge quantities of heat; the 15,000 gallons of water per minute required to cool the plant's radioactive cores come from treated municipal effluent. It's the only nuclear power plant in the world not located next to a river or other large body of water.

More than twenty years' worth of spent nuclear fuel, along with other, less dangerous wastes, have accumulated on the heavily secured six square miles of desert around the plant. The spent fuel—more highly radioactive than the fuel that originally went into the reactor— also has to be kept cool. It sits out there in the desert in deep pools of water or steel canisters because, as Alan Weisman put it in his 2007 book *The World Without Us* (which imagines the fates to be met by the whole range of civilization's artifacts upon the extinction of the human species), "there's nowhere else to take it." He adds that the structure sheltering the plant's atomic swimming pool "was intended to be temporary, not a tomb, and the masonry roof is more like a big-box discount store's than the reactor's pre-stressed containment dome. Such a roof wouldn't last long with a radioactive fire cooking below it."

There was never any way that Phoenix's long-sustained boom could continue indefinitely at the same pace. And as the 2010 census approached, there were indirect indications that the city's population growth might have stopped or even reversed: mortgage foreclosures were at epidemic levels, water hookups were down, trash collectors were taking in smaller hauls, police were getting fewer calls, and sales-tax revenues were down. The *Arizona Republic*'s analysis of these and other trends suggested that "Phoenix has anywhere from a few hundred to several thousand people fewer than projected." Local leaders were wringing their hands. But the city wasn't actually shrinking; it just was not expanding as fast as it had for the previous sixty years. If the population trend actually turned negative, it could give the area's environment a small break, something it hasn't gotten for a very long time.

FLORIDA'S NEW FRONTIER

Two thousand miles southeast of Phoenix, the brand-new town of Ave Maria, Florida, sits at the center of an eight-square-mile tract consisting mostly of bland-looking, semitropical countryside. The small downtown's beige-and-brown skyline is dominated by the sharply arched roof of a one-hundred-foot-tall church known as the Oratory. Facing the Oratory is Ave Maria University, and clustered around the town center are new streets densely lined with spic-and-span houses. Stubs of yet-unbuilt streets lead off in all directions. Founded in 2007, the town of a little more than five hundred residents is situated near the fringe of the Big Cypress Swamp in Collier County, thirty miles northeast as the gull flies from the bustling beaches of Naples. But with its narrow, curving streets, quaint architecture, fresh paint, and (on a warm Tuesday afternoon in December) almost total absence of any human presence, Ave Maria can give a visitor the feeling that he's dropped into a village in some unidentifiable European nation that has suffered an all-out attack by power washers.

Ave Maria is the creation of Domino's Pizza founder and prominent conservative Catholic Tom Monaghan, in alliance with the Barron Collier Company, which was founded in the 1920s by the region's original big-time developer of the same name. During its creation, Ave Maria was the largest single construction site in the country. According to the *Naples Daily News*, "In order to make the property suitable for development, Barron Collier had to carve about 20 percent of the town into retention ponds and use the salvaged dirt to raise the surrounding land between 3½ and 5 feet." In addition to the university, the town plan calls for eleven thousand residences, seven hundred thousand square feet of retail space, four hundred hotel rooms, two schools, and two golf courses. Even as southwest Florida staggered under the impact of recession in 2009, Ave Maria's developers insisted that, within twenty years, the town would have 25,000 residents and the university more than 5,000 students. Much of the controversy surrounding Ave Maria has focused on the question of how it can exist simultaneously as a Catholic enclave and a municipality. But Ave Maria is also an especially striking incarnation of south Florida's most persistent affliction: over-development. It's a problem that has persisted for decades, regardless of the valleys and peaks of the business cycle.

Agribusiness, led by Big Sugar, once posed the greatest threat to south Florida's ecosystems. The long struggle to save the Everglades from agriculture and suburban growth west of Miami and up and down the state's southeast coast is far from over. After some vigorous and contentious restoration efforts in the 1990s, wrote Michael Grunwald in his 2006 book *The Swamp: The Everglades, Florida, and the Politics of Paradise*, "the Everglades were still dying—just a bit slower than before." Now the chief threat to Florida's remaining natural lands comes from residential developments like Ave Maria in the southwest and the tentacles of transportation and commerce that feed them. As the housing bonanza of the early 2000s hit its peak, Grunwald and others lamented the fact that local governments and developers in southwest Florida were creating new environmental crises faster than the older mistakes made on the east side of the state could be corrected. When the national economy toppled, this southwest coast had farther to fall than did most parts of the country. Lee and Collier Counties—which, respectively, contain the coastal cities of Fort Myers and Naples—were hit early and hard as the foreclosure crisis and high unemployment swept through in 2007–08. But the damage done by past overdevelopment won't be easily undone; indeed, further ecological destruction is almost assured, even with slowed housing construction.

The area's transformation has been remarkable. In 1950, according to a history of the Collier Mosquito Control District,

Naples was a sleepy little town surrounded by water and as yet undiscovered by tourists and escapees from the frigid north. . . .

In fact, Naples was hard to find on a map and even harder to get to, at least for people who wanted to visit. Mosquitoes, on the other hand, knew right where Naples was located and they did not need a highway or an airport to be able to reach town. And reach town they did! They severely limited outdoor activities and, in general, made life less than desirable during the mosquito season.

By 1960, as air-conditioning was coming on the scene, Collier County still had a scant 16,000 inhabitants. The population curve then bent sharply upward, reaching around 350,000 today. An additional 150,000 or so tourists converge on the area from December through February. In the boom year of 2007 (the most recent year for which the U.S. Bureau of Economic Analysis has data), Naples had the second-highest per capita income among the nation's metropolitan areas. In 2009, as business and real estate continued to tank, the Associated Press found the wealthy still congregating in Naples and other scenic spots: "Although many Americans are poorer now than at the end of 2007, the geographic distribution of wealth likely hasn't changed much because there have been fewer Americans moving." If anything, well-to-do nonretirees were moving into already-wealthy areas, partly because "the Internet, wireless technology, and the ability to fly commercial in and out of almost any airport in the country have freed [high-income earners] to move elsewhere in significant numbers."

Perennial winter-only residents amount to only 5 percent of the area's year-round population, but their lifestyle would appear to have a bigger environmental footprint per person than that of those who stay put. Seasonal residents may be avoiding the big heating bills they might face if they stayed year-round in their northern homes; however, they are maintaining dual residences, often making more than one round trip per year between them, and often running the air-conditioning in the Florida home during their summer absence in order to protect their possessions from the ravages of humidity.

Writer John Rothchild moved from Miami Beach to Everglades City, south of Naples, in 1973. Affluent migrants from the north—whom Rothchild thought of as the "green and pink people," based on their fashion preferences—were by that time accounting for much of Naples' growth, at least during the winter, when both temperatures and mosquito populations were down. The city's long white beach, he wrote, was "a wonderful beach, supportive of the spirit of the town, the water tepid and placid. . . . There is nothing more lulling than sharing the Naples beach with its Republicans on a windless day." Migrants and visitors, whatever their income level, came to southwest Florida to soak up the warmth, but today, windows stay closed and air conditioners hum year-round; even in winter, a drive-by survey of any Naples neighborhood finds almost all houses with windows sealed and the air-conditioning activated, even on a beautiful 80° day.

As in many upscale areas, sidewalk café–style dining is popular, but most eating and drinking is done indoors, with windows shut. Such a highly mobile population also demands large volumes of space for high-value goods that must not be exposed to the sticky atmosphere. Naples' industrial area features numerous storage facilities, with heavy emphasis on air-conditioned space for items susceptible to heat and damp. That includes air-conditioned car storage: some south Florida owners of high-value vehicles pay $60,000 to $400,000 for "car condos," individually owned, climate-controlled rooms that protect vehicles from the elements. Area residents can drop their cats, dogs, and ferrets off at air-conditioned "pet resorts." Air-conditioned golf carts, pioneered in Arizona and introduced at Orlando's Falcon's Fire course in 2007, will surely catch on in Lee and Collier Counties, which together are home to more than 150 golf courses. The indoor version of that old Florida institution, miniature golf, is also available in Orlando.

Symbolic of the region's environmental predicament are the names chosen for the Florida teams that compete in the world's most heavily refrigerated sport. The minor-league ice hockey team that plays in Lee County's Germain Arena is known as the Florida Everblades, while the National Hockey League team based across the state in Broward County is the Florida Panthers. A sport once considered alien to this semitropical state has taken its mascot names from the ecosystem and the species most threatened by just-as-alien overdevelopment. A more recent project christened after the ecosystem it is helping to destroy—the Big Cypress Swamp—is Naples Big Cypress Market, which lies beyond the city's southern frontier. The complex's centerpiece is an 87,000-square-foot air-conditioned flea market. It also features a miniwinery, performance stage, farmers' market, and tiki bar. Maybe sun and fun once were enough to keep people here happy, but, according to Big Cypress Market's developer, "A lot of people are looking for something other than going to the beach and golfing down here. They go shopping to be entertained to some degree."

"AN ALL-OR-NOTHING THING"

Six million years ago, south Florida lay under a shallow sea. Then the state's original land developers—microscopic marine organisms— went to work. Their calcium-rich remains gradually built up a solid limestone floor, and large parts of that floor rose into the sunshine over the past hundred thousand years as sea levels fluctuated. That water and sunshine have been drawing throngs of new residents to Florida for more than a century, with the bulk of that migration coming in the age of air-conditioning. Florida has gone from being the least populous state in the South to the fourth most populous in the entire country.

Gary Mormino is a professor of history at the University of South Florida, in St. Petersburg, and the author of *Land of Sunshine, State of Dreams: A Social History of Modern Florida.* "It's inconceivable," he told me, "that there would be a Florida of eighteen and a half million people today without air-conditioning." In each of the nearly 150 book promotion events he's done since publishing *Land of Sunshine*, says Mormino, "I make sure to ask the audience this question: what made today's Florida possible? Every time I have asked that question, the first answer I've gotten has been 'air-conditioning.' "

It wasn't only air-conditioning, of course. "No matter how long you've lived in this state—fifty years or one week—you'll know the factors that have created present-day Florida," says Mormino. "There's Walt Disney, Fidel Castro, World War II." Then there has been a series of technological revolutions: insecticides to fend off the vast mosquito populations, automobiles and the interstate highway system to bring tourists, and television and *Miami Vice* to help create the state's image.

"But for all practical purposes, air-conditioning was essential to the development of the Sun Belt in general and Florida in particular. It was unquestionably the most significant factor."

Florida can even lay claim to being the home of air-conditioning. In 1851, Dr. John Gorrie of Apalachicola received a patent on an ice-making machine run by a steam-driven compressor; he suggested that the compressor "could be powered by horse, water, wind-driven sails, or steampower." A physician at the U.S. Marine Hospital, Gorrie used the ice thus produced as the cooling source for an air-conditioning system meant to benefit yellow fever and malaria patients. The state of Florida chose Gorrie as one of its representatives in the U.S. Capitol building's Statuary Hall, where each state is permitted to place statues of two of its most prominent citizens.

Mormino's colleague Raymond Arsenault says that if he were to rewrite his celebrated 1984 article on air-conditioning and Southern culture today, there is little he would need to change. In Florida, the transformation wrought by air-conditioning has, if anything, intensified. For a quarter-century after Arsenault published his paper, cooling technology continued to draw swarms of new people and new ways to the Sunshine State. For one thing, the vast retail jungle that now stretches almost unbroken from St. Petersburg to south of Naples, with its countless square miles of big-box rooftops and parking lots, says Arsenault, "simply could not be the case without air-conditioning, which explains both the demographics and the economy that comes with it. It's sort of an all-or-nothing thing. Naples [where, in some zip codes, the average household spends $35,000 to $40,000 on retail goods annually, far and away the highest rate in Florida] wouldn't be Naples; it would be, maybe, a small fishing village." Arsenault once called the shopping mall "the cathedral of air-conditioned culture," and south Florida's postmall retailing trends—urban pedestrian walkways that combine shopping, eating, and entertainment on the one hand and big-box stores on the other—also depend heavily on climate control.

The spacious, pleasant offices of the Florida studies program led by Arsenault and Mormino are located in a Dutch colonial revival house built in 1904 by early St. Petersburg developer Peter Snell. The house was moved from the Tampa Bay waterfront to the University of South

Florida campus in 1993. It had been built for a hot climate, with big porches and windows, big eaves, French doors, high ceilings, and transoms over interior doors. But today its windows are caulked permanently shut and its air-conditioning system is always on or at the ready. "That really is a shame," says Arsenault. But, he says, if there were no air-conditioning, the heat would probably drive him back to his native Minnesota.

Arsenault is not alone. That is why Florida needs people like Jim Roberts, a longtime air-conditioning contractor who lectures at a local college and once had a call-in radio show on the subject of climate control. Roberts says that at this soggy tail end of the continent, the biggest comfort issue is the humidity, which is more of a problem here than almost anywhere else in the United States. There is plenty of sensible heat (which is what we track as we watch the mercury climb) in southwest Florida; however, the region's humid air is also heavily loaded with latent heat—the energy that has to be removed from water vapor in order to condense it. Wringing water out of air takes a tremendous amount of energy. As Roberts says, "It's not that hard to hold sensible heat down to an acceptable level, but it's very hard to control latent heat." Because of the high moisture content of the warm atmosphere, he says, south Florida has the highest number of indoor "cooling hours" in the entire country, 40 percent more than even Houston, and more than six times what people in upstate New York endure. On Florida's oppressive and ever-expanding roadways, the average vehicle's air conditioner consumes seventy-three gallons of gasoline per year. The only states with higher averages are Arizona, at seventy-six gallons per vehicle, and Hawaii, at ninety-four.

It comes as no surprise when Roberts says, "Today, everybody down here has air-conditioning." By the turn of the millennium, the local press found that neither home builders nor air-conditioning contractors could recall the last time a house without air-conditioning had been built in southwest Florida. It had come to the point that a household without air-conditioning constituted breaking news. In July 2006, a diligent *St. Petersburg Times* reporter managed to track down three local families who were voluntarily sweating out a hot summer with natural ventilation. You'd think a Neanderthal clan had been

discovered alive in the foothills of the Alps. But one of the nonrefriger-
ated home owners, native Floridian John Stewart, insisted that the pe-
culiar lifestyle that he and his wife Sheila lead is in part their response
to the breakneck development that has hit the Tampa Bay area: "There's
no moral issue here. I just see what air-conditioning has done to Florida. How many people would live in Florida if there were no air-
conditioning? Would Pasco County [north of Tampa–St. Petersburg]
have been turned into nothing but bland subdivisions?"

The Stewarts are not alone in living with the heat and humidity of
south Florida. Just eight miles north of shiny new Ave Maria in Collier
County is the very different town of Immokalee. Established a century
before Ave Maria and now situated amid orange groves and fields of
tomatoes and other vegetables, Immokalee is home to twenty thousand people, many of them immigrant agricultural workers. The Coalition of Immokalee Workers has won headline-making victories in
recent years, pushing Subway, Taco Bell, and other fast-food chains to
pay more for the tomatoes that Immokalee workers pick. But those
workers go home each day to trailer parks that present a dramatic contrast to Ave Maria's Euro-style storefronts and mini-McMansions.

Half of all mobile homes in the United States can be found in the
southern half of Florida, and those occupied by well-heeled retirees
can be quite luxurious. But many of those in Immokalee are of an entirely different species. Clustered in sandy lots at the center of town
or scattered around its fringes, the bare aluminum boxes have windows hardly bigger than slits. Clearly nonfunctional air-conditioning
units hang at precarious angles from some windows, while a large
share of the trailers have no air-conditioning at all. The Coalition of
Immokalee Workers maintains that the trailers are unfit for human
habitation.

WATER EVERYWHERE . . .

In south Florida, as in Arizona, it seems that every struggle, sooner or
later, comes down to a matter of water. The lower peninsula was once
covered by an extraordinary array of ecosystems collectively known as
the Everglades. The Everglades are sustained now as then by vast sheets

of fresh water that creep continuously and imperceptibly southward toward the Gulf of Mexico. However, over the past century, thanks to encroachment from the coasts and farming in the interior, the Everglades are down to half their original size.

In the state's southwestern counties, the ecosystem under threat is very different from the more easterly "River of Grass" that has been under siege by Miami-area urbanization and sugarcane farming for so many years. The victim here is the Big Cypress Swamp—a vast, flat mosaic of cypress forests, wet prairies, pinelands, and marshes. The swamp's flora and fauna are just as sensitive to disruption of natural water flows and fragmentation of habitat as are those of the grassy tracts farther east.

To date, most attention has been focused on the region's most charismatic inhabitant, the Florida panther. The big cats require large undisturbed ranges (one- to four-hundred square miles per animal), and their numbers have been thinned severely by the spread of subdivisions and commercial centers across the landscape. But the panther's plight is only one highly visible indicator of much broader ecological degradation.

The Florida Coastal and Ocean Coalition, comprising eight environmental groups, summed up the impact that lax regulation has had on the state's landscape:

An in-depth analysis of satellite imagery by the *St. Petersburg Times* shows Florida has lost 84,000 acres of wetlands to development since 1990. The U.S. Army Corps of Engineers approves more permits to destroy wetlands in Florida than in any other state. Between 1999 and 2003, it approved more than 12,000 wetland permits and rejected just one. The state's permitting rules for wiping out wetlands do not require developers to filter out nutrients, the most common pollutants hurting our waterways. Excess nutrients cause algae blooms and invasive aquatic weed infestations, harming habitat and sea life. The areas of the state that suffer the most from water pollution problems have also lost the most wetlands to urban development. State law discourages regulators from calculating the cumulative toll of issuing thousands of wetland permits every year, even though losing wetlands makes the coast more vulnerable to hurricanes. Without wetlands to filter runoff, Florida's shallow-water aquifers—and thus our drinking water supplies—are at risk.

South Florida leads the nation in water consumption per person—a quantity that doubled between 1950 and 2000. And there are more people than ever using water. Although four hundred people per day moved out of Florida between 1980 and 2000 (most of them, according to Gary Mormino, fed up with runaway growth, sprawl, and water woes), another thousand moved *into* the state each day. Sadly for them, the Florida that people thought they were moving into isn't there anymore. Even the most heroic efforts to square the circle—to keep both the remaining natural lands and the accelerating commercial development of south Florida supplied with clean water—have fallen far short. There simply is not enough to go around. Cynthia Barnett discussed the enigma of chronic water shortages in this water-rich region in her 2007 book *Mirage: Florida and the Vanishing Water of the Eastern United States:* "While farmers use far more water than the general public, growth and development drive the fate of Florida's groundwater." In south Florida's suburbs, keeping the taps running while keeping the streets dry—that is, ensuring that developments don't revert to their natural state as wetlands—is a technological feat performed daily by hydrologists and engineers. "The vast majority of fresh water wasted in South Florida is not really from lawn sprinklers. Rather, it is the amount drained off every day to keep the whole place dry," wrote Barnett.

Back in the 1980s, some scientists began suspecting that you can't manipulate such vast quantities of water and replace so much soggy land with houses and pavement and not affect the local climate. The state's coastal cities and the peninsula's interior had warmed by several degrees on average between 1924 and 2000. The group Environment Florida found that in the period 2000 to 2006, the number of days with highs above 90° had increased significantly. The draining of wetlands thereby generated more business for air-conditioning contractors, but it has aggravated Florida's water headaches. Human-caused changes in the landscape of south Florida directly reduced July–August rainfall by an estimated 11 percent between 1900 and 1993.

In South Florida, nature's ultimate retaliation could come (naturally enough) in the form of water, this time salt water. Under moderate global-warming scenarios, rising seas could flood 15 percent of Lee County and 18 percent of Collier County, including parts of Naples as soon as 2050. In that scenario, says Gary Mormino, "I'd expect that because the threatened coastal areas are some of the most expensive real estate around, the state legislature or Congress would come to their aid with bigger and bigger sea walls—a kind of coastal fortress." But in even more dire scenarios, such as those that foresee the melting of the Greenland ice sheet, no fortress would be large or strong enough. Most of Collier and Lee counties would rejoin the Gulf of Mexico, with some of the higher-elevation subdivisions surviving as neatly paved islands.

Asked about the region's building boom in 2002, Al Hoffman, then-CEO of leading developer WCI Communities, told the *Washington Post*, "There's no power on earth that can stop it!" At the time of Hoffman's prophecy, WCI had done as much as any company to make Naples the second-fastest-growing metropolitan area in the country. Today, Hoffman is gone, WCI is in bankruptcy, and board chair Carl Icahn sold his six million company shares in late 2008 for two cents (not two cents per share, just two cents). Unless or until the seas rise, however, Hoffman will probably be proven right: no power on earth, not even an economic crisis, appears capable of putting a stop to the area's overdevelopment. Naples still has one of the highest per capita incomes in the nation, so there's plenty of money power still in the system to help push roads and subdivisions into new territory. Since the 1950s, when northerners looking for a future home in the sun would shell out the widely advertised "ten dollars down and ten dollars a month" for southwest Florida swampland that might one day be dredged, drained, and built upon, the region's energies have been focused on moving real estate. Writer John Rothchild has put it best: "As Detroit must sell cars, Florida must sell property." Neither cars nor housing plots are selling as well as they once did, but Florida development shows little sign of falling into long-term retreat.

"NOW IT'S SILENT"

Ave Maria and similar communities planned but yet to be built are intended to be anchors of ecologically friendly development in the region. County governments, newly sensitive to environmental concerns, now emphasize elements such as high-density villages, transportation nodes, and mass transit that, they argue, will soften the impact of inserting new, large human populations into the Big Cypress ecosystem. But Conservancy of Southwest Florida spokesperson Nicole Ryan told me that the land and water there simply can't handle the numbers. "We'll have glorified subdivisions leapfrogging one another twenty miles east of Interstate 75 [the former eastern boundary of sprawl]," said Ryan, "and everyone will be driving into Naples."

Novelists, historians, and journalists have exhausted entire thesauruses in their attempts to capture the exuberance with which Florida has welcomed hordes of diverse new residents and all sorts of zany enterprises for over half a century. But—as I asked Ellen Peterson, chair of a southwest Florida Sierra Club chapter—if the environment that originally attracted people to her part of the state is largely gone, why do people keep coming? "A lot of people who live here don't know what it used to be like, so they don't know what they're missing," she said. "Fifteen years ago, just sitting on my porch, I'd hear a symphony of fish every evening, when they were jumping in the Imperial River. Now it's silent, and it's all because of development." The sun and winter warmth are still there, says Peterson, but not much else is left. "Now this area looks like everywhere else. It's all one damn strip mall. One intersection on Bonita Beach Road has pharmacies on three of the four corners! Speculation and greed have ruined it here." And the frontiers of human activity threaten to converge, with new suburbs on the far side of the state elbowing their way into the eastern Everglades, toward the development encroaching from the southwest. As Michael Grunwald put it, "In coming decades . . . south Florida could become an uninterrupted asphalt megalopolis stretching from Naples to Palm Beach. Perhaps it could be called Napalm Beach."

Patty Huff lives in Everglades City, an outpost thirty miles south of Naples in the coastal Ten Thousand Islands area. The town was established by Barron Collier as his original headquarters, but after it was hit by Hurricane Donna in 1960, he and the county government moved to Naples. Says Huff, who moved to town in the 1980s, "It was a company town until around 1960, and Collier owned all the houses. Most of the original houses in the area were built on stilts, both to keep them out of floodwaters during storms and to allow more air circulation in hot weather." Now, with air-conditioning, the stilts are a liability: "The cool air goes right through the floor and out," she says. Today, most houses in the town sit flat on concrete slabs.

Huff says the old houses once achieved excellent air flow. Most were built "shotgun" style, so the front and back could be opened up and the sea breeze could whistle through. "Most of them, like ours, were 'upgraded' in the fifties and sixties, with new kitchens and bathrooms and air-conditioning, and the front and back porches were enclosed. That's what was done with ours well before we moved here. We've been told that originally the kitchen was out on the open back porch, to keep the heat out of the house." Sealing up the houses was as much for protection against mosquitoes as for cooling, says Huff. "I understand that people used to smear their screens with tar to keep out the no-see-ums," those maddening midges that can work their way through normal screens. "Somehow people managed to live down here in those days."

Other architectural features of old Florida houses that helped with cooling were reflective tin roofs, roof vents, double roofs, dormers with windows on three sides, and, later, attic fans. Rita Parker, born in Everglades City in the mid-1930s, was among those who managed a happy life there. "Southwest Florida can be a miserable place," she acknowledges, "but we had a good time, and grew up healthy. We lived outside mostly, where it was a lot cooler than in the house. But we clung to the shade. There was a real small strip of beach at the south end of the island, so we swam, used the garden hose a lot. I got along just fine without it [air-conditioning] until I was in my midforties. I remember going to Miami, into a big department store, and saying, 'Wow! What's this?'"

After that, there was no turning back, and Parker would certainly rather reminisce about the old days than relive them. We were having our conversation on a mild December day, but, she said, "At some point this afternoon, I'll have to close up and put on the air conditioner."

In *Up for Grabs*, John Rothchild (who, with his wife, lived in Everglades City in the 1970s) listed the characteristics that he saw longtime rural Floridians sharing with their neighbors in the heart of the Deep South. Among familiar stereotypes involving pickup trucks and beehive hairdos, Rothchild slipped in a reference to "the acceptance of physical discomfort." I asked Gary Mormino if he thought the people who lived in inland Florida before the age of air-conditioning really were more heat-tolerant. "Think about it," he said. "What was the option? But it's true—it was a tougher generation. Besides, suffering was supposed to be good for the soul. A lot of people think that once air-conditioning came along, there was no looking back." But there was considerable resistance, especially when it came to cooling schools. "You can imagine the debates they had," Mormino said. "The old school board member would say, 'When I was a boy, we sweated and it made us tough, and, besides, air-conditioning costs too much money!' "

Indeed, before air-conditioning, people seemed to revel in Florida's moist warmth. In 1956, at a time when fewer than 4 million people called Florida home and only 10 to 15 percent of the state's households had any type of air-conditioning, the *Tampa Tribune* celebrated the growing popularity of an oversized screened porch that had come to be known as the "Florida room." The article noted that the Florida room was quickly "becoming the center of family living. Families are gathering there for lunch and evening snacks, they're taking guests there (instead of to the living room), and they're making it TV-viewing headquarters. For this time of year and almost year-round in Florida, the out-door life is the desirable one, with blossoming trees and fragrances of neatly kept lawns simply asking to be enjoyed."

In 1959, Hollywood, Florida, home owner Harvey Ford was hauled into court by irate neighbors who said his newly installed secondhand air conditioner was a nuisance in a neighborhood of otherwise open windows, because, said one neighbor, "that infernal noise blasts us out of bed every night." Such conflicts could escalate into an arms race in which Floridians bought their own air conditioners and shut their windows against the noise made by neighbors' units. (Forty-eight years later, the soundproofing was complete. Following a tornado that killed twenty residents of Lake County, some political leaders argued against installing tornado warning sirens in central Florida, partly on the grounds that they "are too difficult to hear inside air-conditioned Florida homes and may confuse residents.") It was also in 1959 that the price of air-conditioning dropped to the point that a builder could cool an entire house for the cost of adding a Florida room, so life turned increasingly indoors in the 1960s and 1970s.

In August 2009, the University of Florida's Bureau of Economic and Business Research reported that Florida, like Arizona, had stopped growing. Based on surveys of residential electric hookups, building permits, and homestead exemptions, the bureau estimated that the state's population had dropped by almost sixty thousand between April 2008 and April 2009. For the first time since the post–World War II contraction of 1946, Florida appeared to be shrinking. Down in the southwest corner, Collier County's population had held steady, but Lee County had shrunk at four times the rate of the state as a whole. By then, the state government had already swung into action. Starting in February 2009, the state senate's Select Committee on Florida's Economy had been busy "streamlining" regulations and relaxing environmental policies. "The overarching economic policy of growth management was approved when Florida was bursting at the seams," committee chair Don Gaetz told the press. "Economic policy ought to be tied to economic reality."

Just as air-conditioning has allowed affluent societies to expand into thermally hostile environments, it is now being looked to as means of extending our current way of life into a thermally hostile future. The goal of this book is not to blame air-conditioning for the many ills with which it is linked or to make you feel guilty for using it. Banning air-conditioning today would be about as popular as Prohibition was in the 1920s, and would do little to bring the deeper environmental, political, economic, and social changes that are needed. Air-conditioning is no longer a just a product of the culture and the economic system; it's an essential component. It was a crucial tool in creating a post-industrial world that now cries out to be transformed. With some forethought, we can hold on to many of the benefits we derive from air-conditioning without the squandering of resources that it now entails.

Undoing some of air-conditioning's harm could require no more than turning switches to "Off," opening windows, and going outdoors. Other climate-control dilemmas are now built so deeply into the structure of society that backing out will be much more difficult. But any energy strategy for the coming decades will be forced to deal with how we handle summer comfort. To ask hard questions about air-conditioning need not raise specters of malaise, poor health, social turmoil, and economic collapse; besides, hazards like those are becoming a bit too familiar already. Turning down or even turning off the flow of refrigerated air could improve our quality of life, but only if even bigger adjustments are made in the wider economy and society. If that can be accomplished, we might find ourselves more relaxed, healthier, less stressed at work, and happier at leisure. Children could have better lives, adults could worry less, and social relations could grow warmer.

2

MAKING THE WEATHER

Air-conditioners in office buildings could collapse under increased heat loads as climate change takes hold and temperatures rise, according to a study carried out at Queensland University of Technology's School of Engineering Systems. Lisa Guan says her computer model of indoor thermal environments and the cooling load imposed on air-conditioners in office buildings shows that most units would not cope under the more extreme circumstances. . . . The results of Dr. Guan's study were released as recent heatwaves in southern Australia sent sales of air-conditioners skyrocketing. . . . "If we do nothing, the cooling capacity of air-conditioners will need to increase by up to 59%," says Dr. Guan. "Obviously this is not feasible."

—The Age, May 6, 2008

It is your human environment that makes climate.

—Mark Twain, *Following the Equator*, 1897

On hot summer days in the cities and towns of southern Europe, heat exhausted from room air conditioners becomes trapped between the multistory buildings that line the region's narrow, picturesque streets. The air-conditioning units raise the temperature of the already-hot outdoor air surrounding them by 10°, forcing compressors and fans to run almost constantly, consuming even more electricity in order to flush heat out of homes and shops. This heat-canyon effect, as it's called, is one of the reasons more Europeans than ever are purchasing their first air conditioners.

A different feedback loop between public and private climate control is playing out on a global scale as air-conditioning contributes to greenhouse emissions and higher summer temperatures, eventually creating greater demand for cooling, which will result in the burning of more fossil fuels and threaten further rounds of warming. In 2008, analysts at Hamburg University of Technology in Germany considered the geographical distribution of cooling and heating requirements under two climate-change scenarios, along with the distribution of human population across the globe, and compared those with the current distribution of climates and population. From this, they projected changes in humanity's heating and cooling demands. Four decades from now, they concluded, the average citizen of Earth will experience 18 to 25 percent less cold weather and 17 to 23 percent more hot weather each year, because populations are growing faster in warm regions of the planet, and all regions will become warmer. Taking population growth into account, cooling demand will rise by 65 to 72 percent. Even though the majority of people now living in the world's hottest climates cannot afford air-conditioning now and probably still won't have access to it in 2050, millions of homes, offices, other buildings, and vehicles on every continent will be newly air-conditioned or be reinforced with beefed-up cooling systems; that will add to energy demand and put greater stress on global efforts to cultivate sources of energy that will not further worsen global warming.

In this arena, the United States is the undisputed champion. Already, air-conditioning is approaching 20 percent of year-round electricity consumption by American homes, the highest percentage in our history. In the commercial sector, it uses 13 percent. Air-conditioning by homes, businesses, and public buildings together was consuming a total of 484 billion kilowatt-hours per year by 2007. Compare this to 1955, when I was born into Georgia's late August heat. That year, the nation consumed a total of 497 billion kilowatt-hours for all uses, not just air-conditioning. We use as much electricity for air-conditioning now as is currently consumed for all purposes by all 930 million residents of the continent of Africa. Imagine that the world's second, fourth, and fifth most populous nations—India, Indonesia, and Brazil—used as much energy per person for air-conditioning as does the United States, the third largest. (That seems only fair. Their citizens see a lot more hot and humid weather than we do.) To do that would require more than 2.4 trillion kilowatt-hours annually—an amount equivalent to all of the electricity used by those three countries, plus Mexico, Italy, the United Kingdom, *and* all the nations of Africa.

Homo sapiens is a tropical species; unlike bison or penguins, we require enclosed, heated spaces in order to survive winters in many temperate climates. In earlier times, fire allowed human migration across the globe. In the industrial world, routine combustion of fuel remains essential to human survival at higher latitudes and elevations, but the heating is often done wastefully. Likewise, while intense heat and humidity can make us truly miserable, we can survive such harsh conditions without air-conditioning except in extraordinary circumstances. Under only the most extreme heat, or when heat stress converges with poor health or other factors, is refrigerated air required for sheer survival. Both academic research and tragic personal stories such as those told in Eric Klinenberg's book *Heat Wave: A Social Autopsy of Disaster in Chicago* show that when heat kills, it's usually under unnatural circumstances of society's own making.

"Psychrometric charts," which plot air temperature against measures of air humidity, are used to outline those regions of environmental space in which we feel comfortable. The American Society of Heating, Refrigeration, and Air-Conditioning Engineers (ASHRAE) publishes psychrometric charts that the building industry can use to design heating and cooling systems. Shaded trapezoidal areas on such charts indicate the combinations of temperature and humidity that have been shown to confer comfort in a calm, enclosed indoor environment. Those little trapezoids covers a very small portion of the vast, warm region of the chart in which humans of sound mind and body can exist and function for long periods in good health, even if not always comfortably. Yes, air-conditioning is remarkably effective in producing comfort, but in all but a very limited set of situations, it is not a biological necessity. Interestingly, the greatest share of resources expended on air-conditioning is not in Phoenix or the Everglades but in those many, much more ordinary places where heat and humidity are not so extreme.

By building a world around air-conditioning, we have made it an essential technology. This is one of the modern world's many positive-feedback loops, in which a system's response to stimuli leads to changes that reinforce that same response, leading to further self-reinforcing changes. For another example, we can turn to Thorstein Veblen, who put the concept in more colorful terms more than a century ago. Acknowledging Plato's observation that necessity is the mother of invention, Veblen added that in a market economy, invention is, in turn, the mother of necessity. The story of air-conditioning stands as one of the purest proofs of Veblen's claim.

CONDITIONING AIR

This book is not a technical treatise on air-conditioning; nevertheless, before examining its influence on America and the planet, we need to take a moment to consider what the term "air-conditioning" is normally taken to mean. Air can be conditioned in many different ways—filtered, heated, cooled, humidified, dehumidified—but here, I am using "air-conditioning" in the usual sense, to refer to a system that uses compressor-driven refrigeration to cool and, if necessary, dehumidify indoor air. There are modifications and alternative methods designed to achieve similar effects, but let's set those aside until the final chapter.

Except in the hottest environments, our bodies are warmer than the air around us. We convert the chemical energy in our food into heat; to maintain a constant body temperature under mild or warm conditions we have to rid ourselves of excess heat by radiating it from our skin and breathing out water vapor from our lungs. We shed more heat, when necessary, by letting the atmosphere evaporate our sweat. The amount of heat that must be unloaded onto our surroundings depends on what we are doing. Table 1 shows the rate at which an average person puts out heat during various activities. As the temperature and humidity of our surroundings increase, we find it progressively harder to rid ourselves of the heat our bodies generate.

Refrigerating the air of an enclosed living or working space is an effective way to help the body shed surplus heat. Refrigeration has its roots in the development of the steam engine two hundred years ago. Some of the heat generated in a steam engine's boiler is converted into mechanical power, which can do useful work that normal ambient heat cannot do. The function of a steam engine—using heat energy to drive a mechanical apparatus—can be reversed, so that we can move heat

Table 1

The human body must rid itself of an amount of heat that is related to body size and metabolic rate, the latter influenced by degree of physical activity. As exertion increases, the body generates more heat, and a greater proportion of that heat is emitted in the form of energy carried by water vapor. These are figures for an average person; greater quantities of heat are typically generated by a larger person. Data are from the American Society of Heating, Refrigeration, and Air-Conditioning Engineers (ASHRAE).

Type of activity	Rate of heat generation (watts)
Seated, at rest	100
Seated, eating	170
Standing, light work	188
Bowling	282
Moderate dancing	375
Heavy factory work	470
Vigorous exercise	530

into or out of a space using mechanical power. In the nineteenth century, Florida's John Gorrie and others harnessed steam engines to run refrigeration equipment, which was first widely used for making ice. Later, in modern air conditioners and refrigerators, electrical energy was used directly to generate cool air. The second law of thermodynamics says that if a system is left to itself, heat will always flow from a higher temperature region or body to a cooler one. With input of energy, refrigeration can make heat flow "uphill," from the cool atmosphere inside a refrigerator or air-conditioned room into the warmer atmosphere outside. Figure 1 is a simple sketch of a conventional central air-conditioning system, showing how it's done.

Notice that the air-conditioning system pictured is using only one kilowatt of electrical energy to expel a quantity of heat energy equivalent to four kilowatts. That does not mean that an air conditioner violates the laws of thermodynamics. The electricity running the unit comes most often from an electric utility, where the quantity of useful energy converted to waste heat—typically by burning coal or natural gas to generate electricity—is greater than the quantity of heat energy removed from a room or house by the air conditioner.

Figure 1

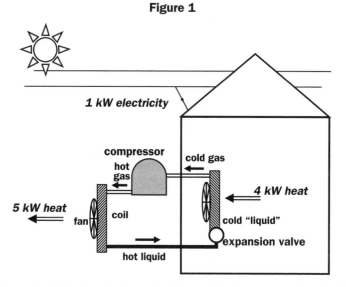

This is a simplified sketch of a central air-conditioning system. See details in the note on page 210.

IT'S HOT AT THE PEAK . . .

For the first half of the twentieth century, air-conditioning's most important role was in manufacturing. It made possible the efficient mass production of all kind of products, including pasta, textiles, chocolate, cigarettes, and photographs. The digital and biotechnological revolutions of more recent decades could never have happened without massive doses of what is known as "process" air-conditioning. But in today's America, seven times as much energy goes into cooling for comfort as goes into process air-conditioning.

Back in 1960, only 12 percent of U.S. households had any form of air-conditioning, and most of that capacity was in the form of window units. By 1980, 55 percent of homes had adopted some form of air-conditioning. But at that point, the industry was just hitting its stride. In 2005, the figure had risen to 82 percent, most of that being central air. Perhaps most remarkably, the average air-conditioned home consumed 37 percent more energy for cooling in 2005 than it had only twelve years earlier. As we will see, part of that increase can be attributed to more widespread adoption of central air-conditioning and part to the growing square footage of houses and apartments. The bottom line: energy consumed by residential air-conditioning almost doubled, from 134 to 261 billion kilowatt-hours, in just twelve years.

Commercial and public buildings together currently use two-thirds as much total energy for cooling as do the nation's residences. During the years between the U.S. government surveys of commercial energy use in 1995 and 2003, the biggest increases in energy for cooling on a percentage basis were in churches and schools, which doubled their consumption. Energy use grew rapidly in retail as well, thanks to the expansion of malls. In those eight short years, total indoor square footage occupied by malls shot up 46 percent, with a whopping 77 percent increase in the energy being used to cool the space.

Electricity use for air-conditioning across all sectors dipped an estimated 2.5 percent between 2007 and 2010 as the general economy shrank. Vacancy rates in the commercial sector rose; however, those rates were still low in absolute terms. Air conditioners were running as hard as ever in the occupied spaces and in some portion of the vacant spaces as well. And the government expects energy consumption for cooling in the commercial sector to leap again by 22 percent over the next twenty years in response to economic growth, despite big projected gains in air-conditioner efficiency. That will put enormous pressure on electric utility companies that are being asked at the same time to reduce their greenhouse emissions.

Peak power demand is the maximum *rate* at which electricity is supplied during a year, a day, or some other precisely defined time period. It should be distinguished from total energy *usage* over time. In the first decades of the electric age, power demand peaked in the winter, largely to illuminate homes during long evenings. Beginning in the 1960s, with greater penetration of air-conditioning, power demand on summer afternoons and evenings began outstripping winter demand in much of the country. By 2007, summertime peak demand exceeded the winter peak by 144,000 megawatts, piling on a quantity of extra demand equal to the output of a couple of hundred large natural gas–fired power plants. The summer peak demand continues to rise gradually but perceptibly year by year; Environmental Protection Agency

(EPA) figures show an increase of 19 percent from 2002 through 2010. Across the warm, humid Deep South, summer peak usage grew by 37 percent over that time.

The industry publication *Transmission and Distribution World* explained that the current system is heavily weighted toward satisfying that peak demand:

> Our electrical infrastructure was built to meet the demands of that mythical July afternoon, which represents something like 10% of the maximum peak demand. Numerous studies have identified this time frame to be . . . about 1% of the total year. We spend a great deal of our resources for that 1% of the time when the aggregate customer's demand exceeds the system's ability to meet that demand.

It's universally agreed that air-conditioning is the biggest factor in pumping up peak loads. Close to half of the increase in demand put on the power grid by California's commercial buildings between early-morning office hours and early-afternoon hours in summer is due to air-conditioning. The Utah Foundation concluded that a growing gap between peak and base demand in Utah is largely attributable to air-conditioning, especially in commercial buildings. In Greece, widespread adoption of air-conditioning, along with heat-island and heat-canyon effects, caused peak power demand to rise by 55 percent during the 1990s.

While coal-fired power plants can supply a base load very consistently, they can't be "turned up" or "turned down" hour by hour to respond to sharp demand fluctuations. That demand is usually met by natural gas–fired plants, which are ideal for giving the needed boost on summer afternoons. Because of that—and because greenhouse gas emissions from gas plants are lower than those from coal plants—the United States almost doubled its capacity to generate electricity from natural gas starting in 2000. Natural gas is generally more expensive than coal to burn, however. Its price is subject to large fluctuations, but an overall upward trend will continue, and probably accelerate, in coming years. Some analysts have warned of a permanent decline in world natural gas extraction as reserves become depleted in the next few decades. But the supply outlook for the United States changed dramatically in recent years, when techniques were developed for extracting natural gas from shale rock. The authoritative Potential Gas Committee estimated in 2009 that the nation's confirmed and potentially accessible gas reserves had grown to more than 2,000 trillion cubic feet as a result, 35 percent above 2006 estimated reserves. A single shale formation—the Marcellus Basin, extending from New York State into West Virginia—may contain 500 trillion cubic feet of gas (in energy content, the equivalent of 80 billion barrels of oil) but it's not known how much of that can be recovered by newly developed horizontal drilling techniques.

Governments and utilities have embraced natural gas as an easy way to meet rising peak power demand, but it also allows them to boast that they are doing something about carbon emissions from coal. But gas is "clean" only in contrast to notoriously dirty coal, just as a bacon cheeseburger can be regarded as healthful if compared with a double bacon cheeseburger. In generating a kilowatt-hour of electricity, it still produces 55 percent of the level of carbon emissions generated by coal.

Other global-warming emissions occur during extraction and distribution. Gas is delivered throughout the United States—not just for electric generation but for space heating, water heating, and other uses—by more than a million miles of pipelines. Because its chief component is methane, a powerful greenhouse gas, the industry has made great efforts in recent years to control leaks from compressors and lines. Leakage has been cut by 19 percent since 1990, but gas continues to escape. In 2007, almost 5 million metric tons of methane leaked into the atmosphere during drilling, pumping, distributing, and other handling of natural gas; that fugitive methane had the global warming potential of 105 million metric tons of carbon dioxide. (By comparison, the nation's 99 million cattle, which are often cited as major greenhouse-gas sources, expel about 6.6 million metric tons of methane from their digestive tracts into the air each year, the equivalent of 138 million metric tons of carbon dioxide.)

Extraction of gas from the earth is a highly toxic business as well. Journalist Abrahm Lustgarten and colleagues have written extensively (more than forty articles for the nonprofit news organization Pro-Publica in 2008 and 2009 alone) to document the environmental consequences of gas drilling. They count more than a thousand cases of contamination that have been reported by courts and state and local governments in Colorado, New Mexico, Alabama, Ohio, and Pennsylvania. Drilling activities allow gas and chemicals to seep into groundwater. Additional contamination occurs aboveground, writes Lustgarten, "where accidental spills and leaky tanks, trucks and waste pits allowed benzene and other chemicals to leach into streams, springs and water wells." However, he continues, "the precise nature and concentrations of the chemicals used by industry are considered trade secrets. Not even the EPA knows exactly what's in the drilling fluids. And that, EPA scientists say, makes it impossible to vouch for the safety of the drilling process or precisely track its effects."

The key innovation being employed to tap newly accessible gas reserves is hydraulic fracturing. In that process, water laced with sand, clay, and chemical additives (known as fracturing fluids) is pumped deep underground to create fissures in the rock and free trapped gas. Most of the polluted water returns to the surface and must be handled as waste. Despite corporate secrecy, some federal, state, and private investigators have managed to identify hundreds of compounds used in fracturing fluids, and many are toxic. Some—including benzene, formaldehyde, 1,4-dioxane, ethylene dioxide, and nickel sulfate—are confirmed carcinogens. Gas companies have enjoyed a slack environmental leash since the 2005 Energy Policy Act exempted them from regulation under the Safe Drinking Water Act and the Water Pollution Control Act; therefore, people living near new or proposed gas-drilling operations receive even less protection than do people who have other kinds of polluters in their neighborhoods.

Drilling in shale can consume hundreds of times more water per well than does drilling in traditional gas fields. In Pennsylvania, which shares the Marcellus formation with four other states, drilling into shale reserves is expected to generate 19 million gallons of waste water daily by 2011, according to the state's Department of Environmental Protection. The water—which carries both natural and human-made toxins and is up to five times as salty as sea water—puts a heavy burden on water treatment plants. The threat to treatment plants isn't limited to chemicals and salts. New York's Department of Environmental Conservation tested waste water from thirteen gas wells in 2009 and found that it was carrying to the surface up to 267 times the safe concentration of radioactive radium-226.

There is no easy way to meet growing summer power demand for cooling while also meeting fuel demand for heating and transportation and still achieve reductions in carbon emissions. That is why the possibility of large new gas reserves has been cause for such good cheer in many quarters. Congressional leaders, financier T. Boone Pickens, and even major environmental groups are hailing natural gas as a high-energy "bridge" to a renewable future or possibly even a "destination" fuel. But the more that is revealed about gas's global-warming and water-polluting potential, the more it looks like a bridge to nowhere.

AND IT'S A LONG WAY DOWN . . .

The air-conditioning of buildings in America is responsible for a quantity of carbon dioxide equivalent to what would be produced if every household in the country bought an additional vehicle and drove it an average 7,000 miles per year. It's one of many reasons that the United States currently leads the world in emissions, both per capita and in total. To shoulder a fair share of the burden in preventing runaway global warming, this country will have to make deep cuts in greenhouse emissions. Straightforward calculations based on current carbon output have led a wide range of researchers—including such respected groups as the United Nations Framework on Climate Change, the Stern Review on the Economics of Climate Change, several European nations, the state of California, and the proponents of the "2000-Watt Society"—to the conclusion that America must reduce its carbon emissions by 80 to 90 percent by 2050. European nations and other wealthy regions will face cuts of 60 percent or more. Even those U.S. political and business leaders in the United States who accept that we will have to cut emissions deeply are loath to recognize that most of that will have to be accomplished by cutting energy use. They prefer more upbeat, vaguely macho language: "energy independence," "clean coal," "green power," "homegrown energy," "carbon trading," "green-collar jobs," or any term containing the word "smart." But, as we will see, renewable energy sources have only a limited capacity to fill the yawning gap between reality and necessity, and none of that "green" energy comes without its own trail of resource exploitation. A deep reduction in emissions requires that great strides be made in energy conservation.

Slashing energy consumption will not only make it much easier to satisfy the remaining demand with renewable generation; it will also help prevent further damage to what's left of the biological world. Serious damage from greenhouse gas emissions lies almost entirely in the future. The environmental damage we see across the globe today was not done by carbon dioxide; it happened through the misuse of abundant, concentrated energy by powerful economic forces. It's not greenhouse gases that have wiped out vast swaths of the South American rain forest, threatened fish populations with extinction, degraded 40 percent or more of the planet's soils, or put more than a hundred previously alien chemicals into the bloodstreams of people and animals in industrial nations. Carbon dioxide didn't pave prime farmland under sprawling suburbs; create the traffic hazards that confront anyone who dares to walk or pedal through American cities and suburbs; threaten the ozone layer, the desert Southwest, and the Everglades; or put millions of homes and farms in India and China at the bottom of reservoirs. All of that damage was made possible by abundant fossil energy. Curtailing the damage will require sweeping economic and political changes, but the job will be more manageable in a lower-energy society.

Existing damage aside, it still will be necessary to dial back electricity use if we are to prevent catastrophic climate change. Most green-power scenarios envision a greater share of energy being supplied by electricity, which already runs almost all air-conditioning. Just under 50 percent of America's electricity comes from coal-fired plants, which have the greatest direct global-warming potential. We have also seen that natural gas is not an ultimate solution to environmental problems either. To some degree, all fuels that generate electricity produce pollution as well. Natural gas and nuclear plants each provide about 20 percent of electrical generation, hydroelectric about 6 percent, and oil 2 percent. Renewable sources account for a little more than 2 percent. All told, America's electric generating plants put out 2.5 billion metric tons of carbon dioxide, 11 million tons of sulfur dioxide, and 4 million tons of nitrous oxides.

Electrical generation from renewable sources is growing fast, but its success is going to mean a long, rough game of catch-up. Generation by geothermal, solid waste, biomass, solar, and wind power combined would not, in 2007, even satisfy the nation's air-conditioning demand; in fact, those sources could expand fivefold and still not produce enough energy to run our air conditioners, let alone serve other uses. The U.S. Department of Energy projects that those renewable energy sources will indeed expand more than fourfold, but not until 2030. By that year, if the department's predictions hold, total electrical generation from all renewable sources will be sufficient to provide only 75 percent of air-conditioning demand and only about one kilowatt-hour out of ten overall. Electricity for all other purposes would still be dependent on fossil fuels, hydroelectric dams, and nuclear reactors.

Every kilowatt of electricity wrung from renewable sources will be hard-won, and air-conditioning will have to compete with society's other top energy guzzlers. The biggest competitor will continue to be transportation. Most scenarios for deep cuts in carbon emissions from cars and light trucks call for greatly expanded use of electric vehicles. If all household vehicles were to be replaced with highly efficient electric models, with no reduction in numbers, the new cars would still consume on the order of twice as much electricity as residential and commercial air-conditioning combined. This might work well from an electric utility's point of view, because the two uses would impose their greatest demands at different times: peak air-conditioning loads would come in afternoons and evenings, while the strongest pull on the grid by electric cars would occur overnight. Yet a large electric-car fleet would require massive growth in the power supply. Electricity from wind or sun would be the most benign way to run personal vehicles, but the small green-electricity sector won't be able to keep up with high-volume users like air-conditioning and transportation, and we will be back to nonrenewable sources.

GLOBAL COOLING

The love of cool continues to proliferate across the planet. As Australian researcher Yolande Strengers put it in 2008, "People are increasing their comfort expectations faster than the climate is changing." Here are a few sightings of recent air-conditioning deployment:

- Between 1997 and 2007, the number of Chinese households owning air-conditioning units tripled, with the annual number sold reaching more than 20 million. Meanwhile, in India, a marketer for the Korean company LG Electronics, the world's largest air conditioner maker, relishes the prospect of middle-class families being driven further indoors: "I see AC sales competing with color TVs as temperatures are going to be a lot worse and as pollution in India is on the rise."
- In Dubai, long famous for its energy excess, builders of the new Palazzo Versace hotel announced in 2008 that it would feature the world's first air-conditioned beach. Coolant pipes would circulate through the sand, and there was a proposal to "install giant blowers to waft a gentle breeze over the beach." The swimming pool's water will be refrigerated as well. In a city that already boasts the world's largest per capita carbon footprint, the hotel chain's president said, "This is the kind of luxury that top people want."
- In the title of his 2000 essay collection, Cherian George dubbed Singapore the "Air-Conditioned Nation." The equatorial city-state built on that reputation in 2005 with the opening of a new night spot. At the Eski Bar on Circular Road, the atmosphere is colder than the drinks. The thermostat is kept at 30° to 31°F, and business is reportedly brisk.
- In Denpasar, on the Indonesian island of Bali, a canine "hotel" features five-foot-by-five-foot air-conditioned rooms. The *Jakarta Post* reports that "there are 32 such units for rent, for Rp 75,000 (US$8.30) per day each. This price includes service and provision of meals. Each compartment has its own bed and does not smell bad."

- When is it too hot for air-conditioning? London's famously chilly, damp weather is giving way more and more to an irritating heat in the summer months. In July 2005, South West Trains, the country's biggest train company, was forced to shut off air-conditioning systems in its compartments because, having been "designed in 1985 to cope with normal British summer weather," the systems could not handle the soaring temperatures. Stuck in the sealed "metal boxes," commuters fumed, but, reported the *Times* of London, "operating a mobile sauna is also contributing to profits as buffet managers report that drinks sales triple on the hottest trains."

- The central role that climate control plays in hot cities can be manifested in many ways. Consider the strategy of punishing inmates at a sprawling south Phoenix law-enforcement complex by denying them air-conditioning. During the area's record-breaking July 2003 heat wave, Associated Press reporter Ananda Shorey visited the Maricopa County Jail, where controversial hard-line sheriff Joe Arpaio "houses" 2,000 inmates outdoors in non-air-conditioned tents and allows them to wear only pink underwear. Inside the tents, "hundreds of men wearing boxers were either curled up on their bunk beds or chatted in the tents, which reached 138 degrees inside the week before. Many were also swathed in wet, pink towels as sweat collected on their chests and dripped down to their pink socks." It is only the human inmates who endure such conditions; an annex to the jail, it turns out, includes an air-conditioned shelter for abused and neglected animals.

- The United Nations Secretariat building, a heat-capturing glass monolith on Manhattan Island, warmed up by 5° in the summer of 2008 when Secretary-General Ban Ki-moon ordered thermostats turned to 77° in an effort to save energy and help reduce the 1960s-vintage structure's carbon footprint. A former ambassador to the UN from Bangladesh— a country that could see its own hot landscape submerged one day thanks to melting polar ice—labeled the move "tokenism." But, he added, "It is important to understand the realities of living in various parts of the world." The secretary-general suggested that delegates and staff, at least those from tropical countries, wear their "national dress" in order to keep comfortable in the warmer temperatures.

Energy-intensive climate control is moving into well-heeled urban and suburban neighborhoods on every continent. In Western nations, every climate-change discussion seems to reach its climax with ominous predictions that automobile and appliance sales in India and China will overwhelm all efforts to curb worldwide carbon emissions. Analysts at the Prayas Energy Group, a Pune, India, nonprofit, acknowledge that "no one would dispute that India and particularly China must rapidly reduce their emissions below projected levels and follow a much less carbon-intensive path." But Western nations, they argue, are using India's and China's future emissions to excuse their own present and past emissions and divert attention from the West's failure even to meet minimal commitments to reverse greenhouse gas emissions. Even if the most restrictive plans put forward by the Organization for Economic Cooperation and Development (OECD) were followed, OECD's economically powerful member nations (which include the United States) would not see real reductions until around 2024; between now and then, they would simply be making up for thirty-five years of overshooting the greenhouse limits they agreed to under the less stringent Kyoto Protocol to the United Nations Framework Convention on Climate Change. (The Kyoto Protocol came into force in 2005 and was designed to hold carbon emissions below their 1990 levels.) The United States did not sign the protocol, but even if it had committed to that modest standard, it would not make up for its own past excess emissions until 2034.

This leaves the international climate standoff unresolved. The U.S.-based activist think-tank EcoEquity, through its "Greenhouse Development Rights Framework," laments that

even among those who see the *climate* crisis, and understand how little time we have to engage it, the *development* crisis still confounds. And since it's the two together that set the unforgiving terms of the international impasse, the question, finally, is if we'll admit that any climate protection framework that remains merely a climate protection framework is doomed to failure, and an ignoble failure at that.

Meanwhile, the finger-pointing goes on between Western nations—which are severely delinquent in reducing their carbon output—and Asia's economic powerhouses, which have big plans to increase their energy consumption. With its larger population and significantly greater affluence than India's, China strikes the greatest fear in Western hearts. But the Chinese government has become impatient with its own citizens' failure to step up and act like world-class consumers. In December 2008, frustrated by statistics showing that Chinese families put more of their income into savings than do the people of any other nation, Beijing expanded a rural program for subsidizing purchases of almost two hundred types of appliances. In February 2009, subsidies were extended to air conditioners along with motorcycles, computers, and water heaters. When people who had never owned an air conditioner showed some reluctance to buy their first one, the government shifted its focus and announced that if current air conditioner owners in any of nine selected cities bought a new unit, the old one would be picked up and "recycled" at no charge, and a 10 percent rebate would be provided on the new purchase. One out of three air conditioners worldwide are already being manufactured in China, many for export, and with the right incentives in the domestic market, production is expected to soar. Meanwhile, as we will see in chapter 7, India's growing middle class needs no incentives to buy vehicles or appliances, and the air conditioner market there is booming.

THE REFRIGERANT PUZZLE

Air-conditioning puts an additional burden on the atmosphere with the unintentional release of refrigerants. Whenever air conditioners and refrigerators are manufactured, transported, charged, operated, or discarded, some refrigerant slips into the atmosphere. The chlorofluorocarbons (CFCs) used in most pre-1990s air-conditioning systems had very high potential for damaging the planet's protective ozone layer. (The breakthrough CFC refrigerant, Freon, was invented in 1930 by chemist Thomas Midgley, working for General Motors' Frigidaire division. Midgley earlier had found that the problem of car engine knocking could be solved by adding lead, which wound up causing serious air pollution and health problems. On the strength of his two momentous discoveries, Midgley was credited by historian J.R. McNeill as having "had more impact on the atmosphere than any other single organism in earth history.") Worldwide alarm at the rapid CFC-induced expansion of "ozone holes" over the polar regions led to the 1987 Montreal Protocol on Substances that Deplete the Ozone Layer, which decreed that CFCs must be phased out in favor of newer, more benign refrigerants in all countries and industries. As a result, the ozone layer appears to be recovering and may return to its original thickness by 2070 or so if we behave well and are lucky.

Hydrochlorofluorocarbons (HCFCs) were seized upon as an interim solution. The most widely used compound for building air-conditioning, HCFC-22, has only 5 percent of the ozone-destruction potential that CFCs have, but international regulators have determined that even that impact is too much. Worse, HCFC-22 has 5,000 times the global warming potential of carbon dioxide. It was banned from new equipment in the United States starting in 2010, and all other wealthy nations will phase it out by 2020. Because of its cheapness and ease of use, China, India, and other emerging economies were allowed to continue using HCFC-22 until 2040. However, the rapidly increasing production of air conditioners and refrigerators in those countries—along with evidence that the Antarctic ozone hole is closing more slowly than expected—has spurred efforts to replace all HCFC-22 in new equipment with more benign compounds such as hydrocarbons as quickly as possible.

To replace the potent ozone destroyer CFC-12 in mobile air-conditioning, automakers turned in the mid-1990s to hydrofluorocarbons (HFCs), primarily HFC-134a. Although HFC-134a does not destroy ozone, one pound of the compound has the global-warming impact of 1,430 pounds of carbon dioxide. As a result, refrigerant leakage from vehicle air conditioners in the United States adds the warming equivalent of 53 million tons of carbon dioxide to the atmosphere annually, on top of about 50 million tons released as a result of increased fuel consumption attributable to air-conditioning. An alternative mobile refrigerant, HFC-152a, would, according to the EPA, have only one-tenth the global-warming capacity of HFC-134a, and, if used in some highly efficient experimental systems, could also reduce the extra fuel consumption imposed on the vehicle by its air conditioner.

By 2000, the world had "banked" 2.3 million metric tons of refrigerants in existing equipment, 63 percent of that in air conditioners, and was producing a half-million tons more each year. Ozone-killing CFCs had been phased out of production but still accounted for one-fourth of the banked supply. Emissions of CFCs dropped slightly during the 1990s but the compounds continued to leak from older equipment through the first decade of the 2000s. Of the planet's total refrigerant emissions in 2000, 37 percent came from cooling buildings and another 26 percent were from car and truck air conditioners.

Whatever their ozone-friendliness rating, all commonly used refrigerants also have significant global-warming potential. Despite the gradual conversion to more benign compounds, market growth meant that the aggregate global-warming potential of leaked refrigerants in 2000 was slightly higher than in 1990. North America, with less than 5 percent of the world's population, holds 43 percent of all refrigerants currently banked inside equipment and is responsible for 38 percent of the resultant global-warming effects. But refrigerant use is growing faster now in other nations, with China in the lead. Propane and other hydrocarbons, ammonia, and even carbon dioxide are among many alternative refrigerants being looked to as replacements for HCFCs and CFCs, because they do not threaten the ozone layer and have relatively low or no global warming potential.

No car, appliance, or chiller plant lasts forever, so a high priority is now placed on destroying first-generation refrigerants from discarded equipment and reusing newer ones, with as little leakage as possible in the process. Because refrigerants are designed to be chemically durable, breaking them down into harmless substances requires sophisticated technology; where resources are lacking, old air conditioners and refrigerators are sure to be dumped and their refrigerant loads allowed to seep out. In *The World Without Us*, Alan Weisman writes:

> If we vanish, millions of CFC and HCFC automobile air conditioners, and millions more domestic and commercial refrigerators, refrigerated trucks, and railroad cars, as well as home and industry air-cooling units, will all finally crack and give up the chlorofluorocarbonated ghost of a 20th-century idea that went very awry. All will rise to the stratosphere, and the convalescing ozone layer will suffer a relapse.

None of the potential replacements for current refrigerants is perfect. Ammonia was replaced in the 1930s by CFCs because it is poisonous and inflammable. And, at least in the case of chillers used in cooling systems for large buildings, replacing HCFCs and HFCs with less efficient refrigerants could result in even *greater* overall greenhouse emissions, according to energy consultant James Calm. Because of their inferior heat-transfer properties, writes Calm, hydrocarbons and ammonia would force cooling systems to use more electricity to achieve the same comfort levels, which means that power plants would be putting even more carbon dioxide and other greenhouse gases into the air. Unfortunately, according to Calm, "there are no known *ideal* refrigerants. Such fluids would offer zero ozone depletion, zero global warming, high [cooling] efficiency, very low toxicity, and no flammability, as well as low cost, high materials compatibility, and high chemical and thermal stability." It is unlikely that any chemical with that combination of traits exists. Compromises are inevitable.

From the 1920s through the 1950s, air-conditioning was a marvel of modern technology, used across America as a marketing tool and to great effect. It still fills that role in newly industrializing countries today. Now it is a given: a house without central air would be virtually unsellable in much of the United States. And a local business that does not have year-round climate control is the rare exception in much of the country. Air-conditioning's role as an essential component of our economic system is now as important as its role as a consumer product. That greatly complicates any effort to achieve comfort, let alone health and prosperity, without accelerating damage to the atmosphere.

There are ways to reduce air conditioners' energy consumption or even to cool without using refrigeration at all; those must wait, however, for this book's final chapter. Before sorting through the alternatives, it is important to examine more closely the ways in which air-conditioning has become so thoroughly integrated into America's economy, society, demography, and even politics. Refrigerated climate control does much more than use energy and emit carbon. We use it to shape our world in many ways, sometimes assigning it the part of hero, sometimes that of villain. It seems to be equally capable of handling either role.

3

THE AIR-CONDITIONED DREAM

On our half acre we can raise enough tomatoes and assassinate enough beetles to satisfy the gardening urge. Or we can put the whole place to lawn. We can have privacy and shade and the changing seasons and also the Joneses next door from whom to borrow a cup of sugar or a stepladder. Few of us expect to be wealthy or world-famous or divorced. What we do expect is to pay off the mortgage and send healthy children to good colleges.

—Phyllis McGinley, "Suburbia: Of Thee I Sing," 1949

These days, I hear people say, "I don't send my kids to their room as punishment anymore. That's where they want to be. It's where they keep in touch with the rest of the world!"

—Christian Warren, associate professor of history, Brooklyn College, 2009

A friend in a neighboring Kansas town told me that on a recent sunny June afternoon, her neighborhood was hit with an unusual power outage. She was in her garden at the time, the only person on the block who was outdoors. She noticed that when power had failed to return after ten or fifteen minutes, neighbors began emerging from their houses. Soon, the yards and streets were full of people my friend almost never sees outdoors in the summer. One woman emerged from her house across the street, pushing her baby in a stroller. This was the first time my friend had seen the several-months-old baby exposed to unprocessed air. The neighbor said, "We were going to drive over to ————'s house, but without power, the garage-door opener doesn't work! Anyway, it's a nice day to walk." As daylight lingered on that longest Saturday of the year, the impromptu block party continued. My friend noticed that no one seemed to have made a move to report the blackout. It was as if the neighbors had tacitly agreed upon an excuse to hang around outdoors a while longer.

America's migration to the suburbs in the last half of the twentieth century was stimulated largely by visions of the outdoor life—by the backyard bliss that urban refugees like Phyllis McGinley expected in the heady days when suburbs were an exciting new phenomenon. Barbecue grills, swing sets, and stepladders remain as familiar complements to the single-family home. But across the Sun Belt and in the new suburbs to which people flee to get away from the old suburbs— the "suburbia of suburbia," to use David Brooks's term—the outdoor life continues to recede as yard space gives way, inexorably, to floor space. On most summer days, the climate-controlled pipeline from house through garage into vehicle to parking garage (beware: potentially unfiltered air!) to office or sporting-goods store and back steers Americans away from the fresh air and elbow room that give the nation's suburban and rural terrains their magnetic appeal. The summer sun is no longer good for much, beyond keeping the local lawn-care company and air-conditioning service in business.

Air-conditioning has shaped American society in many ways, and its central role has received explicit recognition from some weighty sources. A survey conducted at the end of the twentieth century asked members of the Society for American City and Regional Planning History to rank "the top ten influences on the American metropolis of the past 50 years." The results:

1. The 1956 Interstate Highway Act and dominance of the automobile
2. Federal Housing Administration mortgage financing and subdivision regulation
3. Deindustrialization of central cities
4. Downtown redevelopment and public housing
5. The Levittown-style suburban tract house
6. Racial segregation and job discrimination
7. Enclosed shopping malls
8. Sun Belt–style sprawl

9. Air-conditioning

10. Urban riots of the 1960s

None of those influences worked independently of the others. At no. 9, air-conditioning should receive a share of credit for intensifying several of the other influences. As Sun Belt–style sprawl, shopping malls, highway travel, the service economy, and urban redevelopment spread across America, the air-conditioning industry was there to lend a hand at every stage.

When the National Academy of Engineering chose the twenty greatest engineering achievements of the twentieth century, air-conditioning (with its partner, refrigeration) came in at no. 10, just after the telephone and just ahead of highways, spacecraft, and the Internet. But again, a simple ranking doesn't unravel the connections between air-conditioning and the extensive networks of electrification (ranked no. 1) that sustain it, or the modern automobile (no. 2), airplane (no. 3), computer (no. 8), Internet (no. 11), and health technologies (no. 16) that it has helped make possible.

From the 1960s to the early 2000s, through housing booms and busts, air-conditioning and the power to keep it running came within the economic reach of virtually every American household. A comprehensive statistical analysis of the quarter-century ending in 1980 shows how economic forces transformed air-conditioning from a luxury into a mass-produced commodity. The study's author, Jeff Biddle of Michigan State University's economics department, found that escalating incomes throughout the country, along with the falling costs of air-conditioning equipment, did indeed play big roles in the spread of the technology. And, as you'd expect, the lower an electric utility's rates in a given locality or time period (or, equivalently, the higher the energy efficiency of available equipment), the more likely people were to adopt air-conditioning. Biddle found no evidence to support what sociologists call "contagion diffusion"; that is, it does not appear that people were more likely to acquire air-conditioning simply because their neighbors and friends had it. People bought and ran air conditioners not to keep up with the Joneses or because they had seen an especially clever advertisement, but for the simple reason that it felt so wonderful and the price was right. Air-conditioning sells itself. If we can afford it, we install it.

REAL ESTATE'S WEIGHT PROBLEM

In the 1950s, a West Coast counterpart to the Florida room was known as the "California garden": an outdoor space, connected to a house's indoor living space, in which families could find refuge from the heat. In a 2008 history, Gail Cooper wrote that the garden was highly valued on summer evenings because "the indoor temperature of most houses will lag the outdoor temperature by about three hours. Thus the hottest conditions inside the house often come just as the outdoors begins to cool down. With interior spaces hotter than outside conditions, escaping the house was a common strategy." Traditionally, urban dwellers had made their summer evening's escape to public or semipublic spaces like city parks, front stoops, or front porches. After World War II, those who could afford new, single-family houses were looking for greater outdoor privacy. A garden that served as part of the house's living space became, like the Florida room, "a pattern card among middle-class homeowners across the country who embraced the possibilities of outdoor living." The California development company Eicher Homes specialized in integrating the indoor and outdoor spaces of the homes they built, and they often used floor-embedded radiant heating. So "when buyers wanted air-conditioning systems in their Eicher homes as well, the firm was unable to respond; both the aesthetics and the economics of Eicher's modern designs were dependent on radiant heating—that is, upon a ductless house." The design and layout of Eicher homes could accommodate hot-water coils within a concrete floor but not the large system of air ducts required by a central air-conditioning system. And eventually, even in California, adaptation to climate was superseded by climate control. As for the aftermath, Cooper doesn't blame air-conditioning for *causing* "the withdrawal of the family from public community spaces." Rather, "both air-conditioning and the modern garden seem to have common cultural roots in a postwar society that embraced the private and familial."

In those postwar years, the federal government was pushing hard to meet rising demand for mass-produced, affordable housing; the goal was not necessarily to produce elegant, functional landscape architecture. In *Air-Conditioning America*, Cooper explains how the construction and comfort industries responded: "Anxious to incorporate air-conditioning into new residential construction, manufacturers promoted the idea that consumers would save money not only from lower installation costs but also from a thorough redesign of the traditional house." It was much cheaper to allow for equipment space and ducting during construction than to retrofit, so that capacity was included in anticipation of demand for cooling and became a self-fulfilling prophecy. Home plans were stripped of heavier construction materials, movable window sashes, screens, storm windows, large eaves, high ceilings, cross-ventilated designs, and attic fans (the elimination of which also allowed cost-saving reductions in hallways and the pitch of roofs). Shade trees were bulldozed to ease builders' access to the construction site. With central air to keep the house cool, contractors could use lighter, cheaper building materials in smaller quantities while leaving off extra insulation or other energy-conserving features; after all, it wasn't the architects or builders who'd be paying the later utility bills. (Cooper quotes the May 1953 issue of *Fortune* magazine, which described the mass-produced home of the day as a "TV-equipped hotbox.") Air-conditioning was no longer an appliance but rather an integral part of the modern house.

Although a 1950s-era industry executive worried about the problem of "selling the public on the idea that air-conditioning is no longer a luxury," the general public was not the key audience that had to be convinced. Writes Cooper, "Architects, builders and bankers accepted air-conditioning first, and consumers were faced with a fait accompli that they had merely to ratify." In 1957, the Federal Housing Administration began allowing the cost of central cooling to be included in mortgage loans; by 1962, a Florida mortgage company was imposing penalties on houses that did *not* have central air-conditioning because of their potentially lower future value. The Internal Revenue Service even allowed the cost of cooling to be deducted as a health expense in the 1950s and 1960s.

Even after year-round climate control had been incorporated into a large majority of American homes, household energy consumption for air-conditioning continued climbing. One reason: the average square footage of newly built single-family houses doubled during the last half of the twentieth century, and floor space per occupant tripled. In 1950, single-family dwellings were being built with an average of 290 square feet of living space per occupant; by 2003, a family moving into a typical new house had almost 900 square feet per person in which to ramble around. The trend reached its zenith in 2007, when homes exceeding 3,000 square feet accounted for 24 percent of all new construction, compared with 12 percent as recently as 1990. The proportion of all U.S. single-family houses with two or more stories grew to two and half times what it was in 1973. In 1980, 93 percent of new houses had no more than two bathrooms. By 2006, more than one out of every four new houses had at least three bathrooms.

Fewer than half of American single-family houses constructed in 1973 included air-conditioning. By 1990, it was being built into three-fourths of new houses. Today, nine out of ten new houses have it. And climate control inevitably requires more energy when dwellings are larger. On a 100° summer day in Kansas, an energy-efficient 3,000-square-foot house will absorb about 32,000 Btu (British thermal units) worth of heat per hour from the sun; from hot, humid air; and from occupants and appliances. That is the "heat load" that the central air-conditioning system is expected to handle on that afternoon. Sitting down to watch a baseball game on TV, the house's two occupants would need to rid their bodies of only about 1,000 Btu of heat per hour, or about 3 percent of the load on the air-conditioning system. Our central air systems do much less people cooling than they do house cooling. And the bigger the dwelling, the bigger the ratio of house cooling to people cooling.

America's housing and finance industries helped create and, at least for a while, satisfy growing demand for lots of extra climate-controlled living space—to free children from having to share bedrooms, to accommodate Americans' ever-growing bulk of material possessions, to make room for more lavish entertaining, and for myriad other reasons. A surprising 17 percent of all nonrental residences sold annually during the pre-2008 housing boom were personal vacation homes. Sales of vacation houses plummeted with the housing bust, but Americans continue to maintain nearly 7 million second homes—almost 7 percent of all homes, representing enough aggregate surplus living space to accommodate the nation's homeless population twenty times over.

Within a given neighborhood, houses and apartments are sold more or less by the square foot; therefore, in a heavy-breathing house market, the bigger and more expensive the house purchased (or the bigger the extension built onto an older house), the bigger the owner's investment in a lucrative market. Following the late-1990s implosion of the dot-com market, residential housing came to be viewed as one of the most lucrative investments a family could make. Housing wealth came to make up almost two-thirds of the wealth of the median-income U.S. household. That gave housing-related industries a disproportionate degree of influence over the financial life of families, especially those in the middle of the income scale. And the bigger the payday for the builder, the bigger the commission going to the broker, the happier the local furniture and appliance salespeople (beneficiaries of the same easy-credit economy that allowed the low-interest mortgage), and the heavier the load on the local electric utility. McMansion mania was a big factor in spurring a building spree by electric utilities as they raced to increase their generating capacity. Those big houses haven't gone away, and utilities will be required to supply them with power through future booms and busts.

A BIG HOUSE ISN'T GREEN

It has been estimated that 46 million U.S. single-family houses have inadequate insulation, and that if all of them were brought up to International Energy Conservation Code standards, residential energy consumption for heating and cooling combined could be reduced by 15 percent. Most of that nationwide energy reduction would be achieved in northern regions, through lower heating demand. Universally tight insulation would reduce pollutant emissions as well, but here the most dramatic impact per dollar invested would be in the southern states. That's because air-conditioning and electric heat are far more common in the Sun Belt, and generation of electricity, half of which is done with coal, is dirtier than heating with natural gas.

The cost of insulating 46 million homes would be $37 billion, and the net savings on energy expenses over the next fifty years, adjusted for inflation, would be about $80 billion. In 2009, the federal government allocated a big-yet-small $5 billion worth of economic stimulus money to an existing program for weatherization of low-income homes. That was more than the total amount the program had spent over the previous thirty years. At the urging of Sun Belt lawmakers, Congress also doubled the percentage of weatherization program funds designated to be spent on homes in warm climates, to reduce energy wasted by air-conditioning. Some groups objected to the new formula, saying that insulation would be more effective, dollar for dollar, if spent in the northern tier of states to reduce heating costs. Compared with the bailout sums that the government was throwing at American corporations in 2008–09 to address the financial crisis, the entire $37 billion required to help address the looming climate crisis through universal insulation should also have been within economic reach.

In areas with cold winters, high energy efficiency requires "tight" houses with good insulation. Those characteristics are needed in hot weather as well, assuming that the air-conditioning will be either running or on call around the clock. But it is neither simple nor cheap to combine the tightness called for by heating and central air-conditioning with the extensive natural ventilation that may be desirable when temperatures are high but not intolerable. Across America, especially in areas where quick fluctuations in temperature are common, houses are staying zipped up tight right through the spring, summer, and fall, as snugly closed when the outdoor temperature stands at 78° as when it hits 99°. The more well-fitted a house is for indoor climate control, the less likely it is that occupants will go to the trouble of opening it to the environment on mild days or nights only to shut everything up again within a day or so. The tendency is reinforced by the desire to keep homes clean. It isn't just in dust-plagued locales like Phoenix that windows are kept sealed year-round to shut out dirt and pollution, and the result can be a significant improvement in quality of life. A tight barrier between the indoors and outdoors is maintained partly for health concerns, as we will see in chapter 6, but it's also an effective way to reduce the burden of housework.

The triumph of central over room air-conditioning in the 1950s and 1960s was a thoroughly American phenomenon, and it is another major reason that we expend so much energy on comfort. For residential cooling in most other countries, room air-conditioning is the rule. There, equipment is often confined to one or two parts of the house and is turned on only intermittently, when the weather calls for it, and only when a space is occupied. In most of the world, people cooling, not space cooling, is the priority. Researchers studying residential airconditioning use in Japan in the early 1990s found that people cooling was universal, and that many residents believed that "cooling or heating of rooms while not occupied is wasteful and unwise." In India as well, middle-class families today will gather around the air conditioner on summer evenings in the same way people will gather around the fireplace on cold winter nights in temperate regions.

As the square footage of the average American residence grew, the construction savings made possible by air-conditioning were eroded. And typically, only about one-tenth of a house's total energy consumption occurs while it is being built. The other 90 percent happens while it's being lived in, and standard, cheap building materials generally hurt energy efficiency over time. More costly "green" construction helps, but excessive square footage can also cancel out improvement via that route. A 2005 study published in the *Journal of Industrial Ecology* arrived at the estimates in Table 2 for energy consumption by three kinds of houses: large with good energy efficiency, small and efficient, and small and inefficient.

The authors concluded that "a 1,500-square-foot house with mediocre energy-performance standards will use far less energy for heating and cooling than a 3,000-square-foot house of comparable geometry with much better energy detailing." Note the important word "geometry." To avoid outsized suburban manors looking like big-box stores, builders have loaded up their product with multiple rooflines and gables, dormers, bay windows, and other protuberances. That

Table 2

Energy required for heating and cooling in two U.S. cities is greater for an energy-efficient large house than for a small house, regardless of the small house's efficiency level. (Model-derived data are from BuildingGreen, Brattleboro, Vt.)

City	House size (sq. ft.)	Energy efficiency	Energy required to heat and cool (million Btu)
Boston	3,000	Good	92
	1,500	Poor	60
	1,500	Good	48
St. Louis	3,000	Good	90
	1,500	Poor	61
	1,500	Good	49

would be excellent for cross-ventilation if the windows were ever opened. But such houses have more surface area than does a squared-off house of the same size, putting a bigger burden on cooling and heating systems. Additional energy is wasted by longer ducts and hot-water pipes. For a given house design—"green" or standard, monolithic or pseudo-Victorian—the bigger its square footage, the bigger its environmental footprint.

Wherever it lands, the big-house trend pushes up consumption. According to recent reports from Australia, where air conditioner sales are growing at a 10 percent annual rate, "peak demand was higher in so-called 'energy efficient' housing developments than in normal housing developments. Many blamed this on the increased size of housing, the fashionable open-plan format, and central air-conditioning and heating." In the state of Victoria, a 30 percent increase in air-conditioned floor space has wiped out energy savings from a new, compulsory housing efficiency standard.

The prospect of resource savings might simply provide real estate agents with yet another pitch to induce home buyers to purchase more square footage; the buyer need only be told, "Hey, with this big, efficient house, you'll get a couple more rooms and you'll be heating and cooling them practically for free!" The old discount-store slogan "The more you spend, the more you save!" applies to home energy conservation as well. So, despite consuming far more total energy, the owner of a big, green house can boast of much bigger energy savings than the small-house owner. Green home builders' sales targets will tend to be more affluent buyers who can afford the more expensive materials and labor-intensive construction necessary to build nominally ecofriendly houses. Many such affluent buyers want their houses big.

Through the housing boom, the phalanxes of jumbo houses that lined suburban streets were seen as protection against erosion of property values. A late-boom survey found that twenty-nine "McMansion communities" had mean incomes 48 percent higher than those of the metropolitan areas in which they were located. The study's authors asked city planners why people built McMansions in their communities, and the most common response by far was "increased land values." Were a house in the midst of such a community to feature a summer comfort system other than central air, it could easily drive down neighbors' property values. I saw this borne out when I came across an early 2006 discussion on AppraisersForum.com ("the premiere online community for discussion of real estate appraisal"). A home owner unconnected to the appraisal profession joined the discussion in order to ask how more ecologically sound building practices and resource efficiency are taken into account in valuing a smaller house that is surrounded by bigger ones. He wrote, "I have to ask because we are building better vs. bigger. By doing so, our house will be about half the size of what can be bought with the same price in our area. But I can't help but think about how it will be appraised if we ever need to sell it." In the collective professional opinion of the appraisers who responded, the home owner was out of luck if he was hoping to have his house valued for its quality and efficiency. Typical of the comments were that "this house would have much 'value in use' to you . . . but, it would be tough for an appraiser to quantify" and that, having spent more money to build fewer square feet, the home owner would "likely pay the price for it at selling time."

Another roadblock to resource conservation efforts is erected by home owner associations. Twenty percent of Americans now live in homes subject to rules set by these private, government-like bodies.

They have sweeping powers to dictate almost any aspect of a member's property, from the size of the residence down to changes in trim color or placement of a basketball hoop. The more restrictive groups cling to outdated standards that treat necessary features of a more resilient future society—renewable energy devices, clotheslines, vegetable gardens, fruit trees, compost bins, natural landscaping—as eyesores to be buried under impossible restrictions or banned outright. The "covenants" that associations impose on home owners not only tend to prohibit green practices; they also mandate overconsumption. Fans in front windows are outlawed in some communities, while others prohibit awnings or window air-conditioning units that are visible from the street. Often, and especially in Sun Belt states, covenants dictate a minimum of air-conditioned floor space for new houses. It is not uncommon to see requirements for 2,500 to 3,000 square feet of air-conditioned space.

By the time the real estate bubble burst, a few local governments had begun trying to slow the malignant growth of houses in their jurisdictions. A search of major newspaper archives between 1998 and 2004 found that forty U.S. communities had adopted anti-McMansion policies and fourteen states and thirty-three other communities had considered them. Most such moves had happened on the coasts or in and around big cities. The bloating of new houses continued right through the first stages of the national implosion. Then, by the summer of 2007, median floor space finally shrank slightly, by a marginal sixty square feet per house. But builders, buyers, and real estate brokers have taken square footage to such heights that hard times alone won't bring it down to a reasonable level. The chief marketing officer for Toll Brothers Inc., the largest builder of luxury homes in the United States, conceded after the bust that there "probably is more demand for 3,000- versus 6,000-square-foot" houses—a seemingly big shift, but only at the very top end. The National Association of Home Builders made a postbust prediction that the average size of new houses, which had grown from 2,350 square feet in 2004 to a little over 2,500 square feet by 2007, would hover at "about 2,300 to 2,500 square feet in 2015." With the deep drop in house prices, those who could still afford to buy would be able to get more floor space for their money. With the market downturn, according to a Las Vegas builder, "We're finding that people are choosing a bigger box for themselves and their families rather than a more luxurious home." Between that average square footage in new construction—still far out of line with other nations, whether poor or affluent—and the millions of oversized houses left in the wake of the real estate rush, the burden on energy supplies will go undiminished.

Virtually all of those big-box houses continue to be heated and cooled. Despite the apocalyptic tone of news reports, the housing-market meltdown did not leave a huge percentage of dwellings standing empty nationwide. Between 2006 and 2008, the national vacancy rate rose from 1.3 to 1.4 percent. Ninety-eight percent or more of the oversized houses built over the past couple of decades will be occupied at any given time during the next few decades, so it will become a matter of public policy to relieve the burden on the nation's energy resources and the planet's atmosphere. If economic necessity intervenes, we might see larger numbers of occupants per housing unit, with square footage per person reversing its historic climb. But whatever the real estate roller coaster has in store for us in coming years, house size will have to be separated from the tangle of resource issues and dealt with as a serious problem in its own right.

SEVENTY-TWO DEGREES OF SEPARATION

In Luis Buñuel's surrealistic 1962 film *The Exterminating Angel*, a group of dinner-party guests find themselves trapped together for days, unable to leave the hosts' dining room even though there is no visible impediment to their departure. The results are not pretty. Their mysterious immobility goes unexplained in the film, but the ties that hold people indoors in the real world of today are no mystery. The list of products capable of immobilizing able-bodied people indoors grows by the month, but for a good part of the year in most of the country, air-conditioning keeps an especially tight grip.

Participation in sixteen different types of outdoor activities, such as visiting public lands and parks or hiking, has declined 18 to 25 percent since the 1980s in the United States, Japan, and Spain. Swimming, picnicking, hiking, and softball remain popular in America, and yards must be maintained, but casual observation shows that time spent outdoors in the typical neighborhood has dwindled. It may be true that "a perfect summer day is when the sun is shining, the breeze is blowing, the birds are singing, and the lawn mower is broken," but such days can pass largely unnoticed in the age of air-conditioning. In my Kansas neighborhood, a shady suburban street on a pleasant 85° summer evening can be as free of human life as it would be on Super Bowl Sunday with a sleet storm in progress. The South led the nation into the age of air-conditioning, and there, Ray Arsenault stressed back in 1984, "human interaction with the natural environment has decreased significantly since the advent of air-conditioning. To confirm this point, one has only to walk down almost any Southern street on a hot summer afternoon, listen to the whir of compressors, and look in vain for open windows or human faces." Keeping us indoors and comfortable, air-conditioning reinforces our already-tight focus on the individual or nuclear family rather than on a larger community, in a trend that Robert Fishman called the "cocooning of America."

"I am amazed at how quickly air-conditioning has become a part of human nature," says Christian Warren, an associate professor at Brooklyn College who specializes in the history of U.S. public health and medicine. He has dug deeply into the questions of how and why life has moved indoors. Although it's not an entirely new phenomenon—it started more than 150 years ago in this country, he says—the trend accelerated in the late 1970s and early 1980s. "Something big happened during that period, something we still don't fully understand," Warren told me. "But we do know that efficient heat pumps, better wall and ceiling insulation, improved windows . . . all of these conspired to make climate control more affordable, allowing houses to get bigger, more comfortable, and more attractive—in the sense of curb appeal but also in the sense of attracting people in from the outdoors." And, he says, advances in other areas such as home food and movie delivery and entertainment and communications technologies have pulled people of all ages indoors, leaving a gap in life experience: "Time spent in the unstructured outdoors is very different from watching the National Geographic Channel. Simply learning about the environment in the abstract makes children detached and fatalistic."

The great outdoors has lost the summertime advantage it once enjoyed. In the temperate zones, winter has always extracted a price from those who would spend leisure time outdoors, demanding that we trade away some of our thermal comfort in order to enjoy the pleasures and risks of the wider world. Air-conditioning has presented us with a similar trade-off in the warmer months as well. Its effects have been reinforced by a profusion of legal, economic, and social barriers between us and the natural world. Those barriers were examined by Richard Louv in his 2005 book *Last Child in the Woods: Saving Our Children from Nature-Deficit Disorder*. Although those barriers can be overcome without much difficulty, Louv is not alone in his concern that we have become alienated from nature; a host of other writers and researchers have expressed concern about our separation from nature in recent decades. In summer, prime time for children to throw themselves into free exploration, the temperature difference between a sunny meadow and a cool, dry family room can easily tip the balance toward the couch. Even the shopping mall is losing its allure. It is no longer the primary indoor venue where young people go to meet, says Warren, because now they can "meet" online. "In my community, which is suburban, kids aren't even learning to drive. And although they all have bikes, they hardly ever ride them. They don't feel much need to go anywhere."

In his recent synthesis, and in light of the current revival of environmental awareness, Louv wrote, "A kid today can tell you about the Amazon rain forest—but not about the last time he or she explored the woods in solitude or lay in a field listening to the wind and watching the clouds move." Despite a flurry of recent research in this area, hard numbers remain sparse; it's not even known precisely how much time children once spent in the "unstructured outdoors" or how much they spend now. But the bulk of data collected so far are consistent with a general suspicion that nature is mostly excluded from most kids' lives. Louv linked cause with effect to the extent they can be linked at this point. Children are separated from nature, he wrote, by diverse, not always obvious forces: the migration from farms to cities; the growing distances that must be traversed between home and open spaces; the burden of homework that grows heavier by the year; those organized activities outside of school that have relentlessly sopped up spare hours and minutes; the longer working hours faced by parents; the media-fed fear that all woods and fields are teeming with child snatchers; corporations' interest in keeping young eyes on their products rather than on trees and insects; concerns over legal liability if children suffer harm in a natural setting; and local ordinances and neighborhood-association rules that ban activities such as tree climbing that were once considered normal.

The consequences of nature-deficit disorder, Louv believes, are serious. There are, as we will see in coming chapters, only partially answered questions about physical health. But it is children's mental health that may suffer the most. Contact with nature and the unregimented outdoors appears to be a powerful stress reliever, and it can expand a child's network of friendships even while offering opportunities for solitude. On the other hand, research shows that green vegetation within a child's routine field of vision can promote mental and emotional health. Recent studies suggest that unorganized time spent in undeveloped green spaces pumps up children's creativity and their ability to focus attention, whereas indoor activities tend to increase the symptoms of attention-deficit disorders. Louv acknowledges that research in this area has a long way to go but argues that all signs are pointing in the same direction. He adds, "Even the most extensive research is unlikely to capture the full benefits of direct, natural experience."

Air-conditioning has helped pave the way for educational initiatives—including mandatory summer school, year-round school, and abolition of recess—that add even more hours to the time children spend indoors annually. Results of a national survey published in 2009 show that about 30 percent of eight- and nine-year-olds in the United States now have no recess or a severely limited version, despite research showing that recess improves attentiveness and behavior in the classroom. Recess advocates also claim that it offers students a crucial opportunity simply to get outdoors and interact spontaneously. Antirecess forces have responded that a lot of the so-called spontaneous interaction comes in the form of fighting and bullying, which can be better controlled if kids are kept off the playground. But bad behavior isn't *caused* by those few minutes outdoors, and the roots of the problem aren't addressed by banning recess. Indeed, lack of exercise and too little exposure to the outdoors could be doing a lot more to foster aggressiveness and lack of focus. As *Time* magazine put it in a 2001 article, "Good-bye recess, hello Ritalin."

Some education "reformers" have also been targeting the summer vacation from school as a relic of the pre-air-conditioning era, and therefore a practice to be discarded in favor of a year-round schedule. The National Association for Year-Round Education (NAYRE) has members from across the political spectrum, but business interests have been prominent in the push for year-round school. Frederick Hess of the right-wing American Enterprise Institute wrote in 2006, "Summer vacation once made good sense—back when we lived in a brawn-based economy, academic achievement mattered less, an absence of air-conditioning or modern hygiene turned crowded schools into health risks, and children had moms who were home every day." Today, he argued, those lazy summer months only provide more time for kids to go unsupervised, get into trouble, and forget what they learned over the school year (unless, he claims, they attend summer camp). The problem, according to Hess, is most serious in low-income families, and that is a serious threat in a world where "our children will find themselves competing with peers from Europe, India, and China for lucrative and rewarding brain-based jobs." In his world, air-conditioning could allow those kids who require more control to sit out much of the summer in comfortable classrooms. But, Hess adds, "Summer vacations are still a wonderful time for many families and communities. Legislators need not pursue one-size-fits-all solutions to 'fix' the school calendar." In other words, in upscale communities where parents can afford to organize their children's summers to Hess's standards, society can afford a long school vacation. Less privileged kids will find themselves in a more controlled July climate. Critics derided what they saw as the "bottom line" for Hess and other year-round-school advocates, with observations like this: "There is no time for leisure time in the dog-eat-dog global economy the business elite has mapped out in its cradle-to-grave plans for America's 'human capital'—their label for our children." The vision promoted by Hess and NAYRE remains deeply unpopular, with only about 5 percent of children subject to year-round school schedules.

The very real need for supervision of young people in an increasingly perilous world has, inevitably, restricted recreational opportunities. In lamenting the disappearance of outdoor play, Walter Kirn and Wendy Cole observed in *Time*, "After all, play needs to happen somewhere—preferably somewhere safe and open and not entirely dominated by grown-ups—but those idyllic somewheres are growing scarce." They quote Hofstra University professor Rhonda Clements: "In the huge rush to build shopping malls and banks, no one is thinking about where kids can play. That doesn't generate tax revenue." Kirn and Cole continue:

> One place kids keep rushing to is Chuck E. Cheese, the chain of video game–crammed pizzerias where families can frolic in air-conditioned safety, separated by turnstiles from the Big Bad Wolf. Such enterprises fill the play vacuum with something far more modern and secure—"edutainment." It's a growing industry. Randy White is CEO of White Hutchinson Leisure & Learning Group in Kansas City, Mo. His company develops cavernous play facilities, up to 30,000 sq. ft. in area, that are Xanadus of prefabricated diversion, offering art projects, costumes, blocks and even simulated fishing. "We're reintroducing free play to families," says White. Free play at a price, that is. His facilities charge up to $10 a head. "Parents feel that if they're not paying much for an experience, it's not worth it educationally," he says.

That was back in 2001. Recent events at Chuck E. Cheese's franchises would doubtless puzzle those early-twentieth-century academics who once argued for the civilizing effects of a regulated indoor climate. Chuck E. Cheese's restaurants have started receiving more press coverage for hosting brawls than for serving pizza. According to the *Wall Street Journal*, police were called to break up twelve fights, all involving adults, at the chain's Brookfield, Wisconsin, restaurant in 2007–08 alone. One melee involved "as many as 40 people knocking over chairs and yelling in front of the restaurant's music stage, where a robotic singing chicken and the chain's namesake mouse perform." In Bakersfield, California, police are called to disturbances at the local Chuck E. Cheese's franchise more often than they're called to the Déjà Vu Show Bar strip club.

David Nicholson-Lord, author of the book *Green Cities—And Why We Need Them*, sees a massive burden of social, psychological, and physical problems now being carried by human populations who have come to live mostly indoors, in the modern equivalent of our prehistoric ancestors' caves. The result of that almost exclusively indoor life, he writes, is "maladaptation on a grand scale—a species moving to a habitat that does not suit it." The remedy is obvious, according to Nicholson-Lord: "We really do need to get out more."

4

GOING MOBILE

The capital of the new planet—the one, I mean, which will kill itself off—is Detroit.

—Henry Miller, *The Air-Conditioned Nightmare*, 1945

Long-distance mobility is an enduring feature of the U.S. population; in a typical year between 1947 and 2005, 2.5 to 3.5 percent of the nation's residents moved to another state. And the movement has not been random. Following World War II, the rapid spread of two technologies—the automobile and air-conditioning—dramatically altered the population distribution of the United States. The car provided unfettered mobility, and residential, commercial, and mobile air-conditioning helped steer migration toward the nation's warmer regions. The populations of booming Sun Belt cities and suburbs, in turn, have found themselves more dependent on both their personal vehicles and, of course, their air conditioners, than are the populations of most northern metropolitan areas.

HEADING SOUTH

During the first half of the twentieth century, the economically depressed southern United States lost well over ten million people. By 1950, the string of nine coastal Sun Belt states from Virginia to Texas, plus New Mexico, Arizona, and Nevada, had a combined population of only 33 million, less than half the total population of fifteen New England and Rust Belt states that stretch from Maine to Minnesota. The 1960s brought a dramatic turnaround, with more people moving into the South than out of it. By 2002, the population of those twelve Sun Belt states had doubled and then grown by another third, to 88 million—almost as many people as then lived in those fifteen northern states. The lines tracing the South's adoption of air-conditioning and the growth of north-to-south migration ran in parallel. That does not prove cause and effect, but neither would their common trajectory appear to be a matter of pure chance.

The scale of Sun Belt growth cannot be fully accounted for by the vast retirement communities of Florida and Arizona; rather, the twenty-five-to-thirty-four-year-old age group moved in much larger numbers, bringing young children with them. Mild southern winters have always been a big attraction. But early immigrants to the South had to pay a price in summer discomfort. As an annual average, residents of those twelve Southern and Southwestern states experience three times as much weather calling for air-conditioning as do people living in the fifteen Northern states. All of the hotter Sun Belt cities gained population during the age of air-conditioning (Atlanta the least, by 26 percent; Las Vegas the most, by a dizzying 1,843 percent), while all but one of the cooler northern cities lost residents, with losses ranging from 8 percent for Milwaukee to more than 50 percent for Detroit and Cleveland; New York was the one northern gainer, up 3 percent.

The Census Bureau allocates the fifty states to four regions: the Northeast, the Midwest, the South, and the West. By 1993, 88 percent of homes in the South had air-conditioning, and that region was responsible for 69 percent of all electricity used for residential cooling nationwide. Over the following twelve years, the number of air-conditioned homes in the South grew by another third, but the most rapid adoption occurred in other regions (see Table 3). In the West, electricity consumption for cooling the average home almost doubled. That had much to do with the rapid growth of Phoenix, Las Vegas, and inland areas of Southern California.

White people were prominent among the retirees who flocked to Florida, Texas, Arizona, and other states in search of naturally warm winters and climate-controlled summers. A study comparing migrants aged sixty and over who moved from the Snow Belt to the Sun Belt in the late 1960s with over-sixty residents who had been living down south all along found some big economic differences:

While there was no significant difference between the two groups on job income, migrants had significantly higher income on all other measures (except welfare income). Migrants had more than twice the income from other sources—pensions and returns from investments. . . . Individual [total] income for migrants was about one-third higher than for the residents. . . . [T]here were somewhat less than a third as many blacks among the migrants as among the resident population.

Table 3

Between 1993 and 2005 in the United States, electricity consumption by residential air-conditioning doubled or tripled in every region of the country, and the quantity of energy used for cooling by the average air-conditioned household rose by one-third, on average. Data from Energy Information Administration, 1993 and 2005

Region	Total energy consumed by residential air-conditioning (billion kilowatt-hours)		Energy used annually, on average, for cooling an air-conditioned household (kilowatt hours)	
	1993	**2005**	**1993**	**2005**
Northeast	12	22	1,073	1,333
Midwest	22	43	1,427	1,838
South	91	166	3,078	4,300
West	9	30	1,222	2,230
United States	134	261	2,113	2,797

Migrants also tended to be home owners, paying out property taxes while adding "little pressure to local public institutions such as schools."

Retirement migration had a multiplier effect, drawing even larger populations of younger people into the South. The swelling population of well-to-do elders increased demand for commercial services in retail, health care, food service, entertainment, and other sectors, and most of the older migrants did not compete for the jobs that their buying power helped to create. It is calculated that the older migrants who moved to Florida just during the years 1985 through 1990 infused the state's economy with an additional $4.9 billion, and that led to a $12 billion increase in industrial output and 160,000 new jobs. One analyst wrote, "Much of the nonretirement migration may be caused by the same amenities"—including comfortable weather—"that cause retirement migration, but a great deal of nonretirement migration also appears to be in response to growing job opportunities." Another noted that as migration to Arizona accelerated in the 1990s, "the majority of new arrivals in the Phoenix area . . . are young married couples with children seeking good jobs, pleasant weather, and affordable single-family housing." The economic effect was self-reinforcing as manufacturing companies were drawn south, seeking proximity to markets. Large and small companies were further attracted by the fact that the bulk of those Sun Belt jobs were nonunion.

The shift of largely white populations toward warmer climates sharpened a racial redistribution that began a century ago as many descendants of former slaves abandoned the South and its continuing racial oppression. Between 1910 and 1950, the North's black population swelled by 12 million, an increase made up mostly of people born in the South and their children. After 1950, a small number of those northern black families joined in the rising tide of migration to the South. On arrival, they found a region divided along thermal as well as racial lines. Ray Arsenault notes that by 1960, 18 percent of Southern households had air-conditioning but only 4 percent of black Southern households had it; a decade later, adoption had reached 50 percent overall but only 21 percent for black households. Since then, air-conditioning has become close to universal and therefore is no longer much of an indicator of a racial divide. (However, we should not forget people like the immigrant workers living in those overheated trailers in Immokalee, Florida.)

Domestic migration has radically changed the nation's political map as well: In the wake of the 2004 presidential election, Hofstra University professor James Wiley penned an op-ed column titled "Blame Air-Conditioning for Kerry Loss." The headline overstates the case, of course, and in the column, Wiley recognized that air-conditioning was only one of several factors behind the political rise of the South. But the southward migration that air-conditioning helped make possible has shifted the country's political center of gravity.

With strong showings in every U.S. census from 1960 to 2000, the Sun Belt realized a net gain of eighty-six seats over the New England/ Rust Belt group of states in both the U.S. House of Representatives and the Electoral College. The South's growing power in Washington during the age of air-conditioning was paralleled by the region's growing support of conservative politics and the Republican Party. Taken together, those trends—along with the winner-take-all method by which most states award their presidential electoral votes—could indeed have tipped the presidential elections of 2000 and 2004 toward the GOP. If we could travel back to 2000 and have each state vote red or blue just as it did that year but with relative populations and electoral votes distributed among states as they had been in the 1950s (before the big southward migration), Democrat Al Gore would defeat Republican George W. Bush by eighteen electoral votes instead of losing by three. It's safe to assume that, had that happened, there would never have been a George Bush–John Kerry matchup in 2004. Nevertheless, if we run that election with the red-blue pattern of 2004 but a pre-air-conditioning electoral vote distribution among the states, Kerry finds himself at the end of election night down by only two votes instead of thirty-four. That almost certainly would have led to recounts in Iowa and New Mexico, where Bush squeaked by with a fraction of a percentage point. (In 2008, incidentally, Barack Obama would have won regardless of whether the electoral vote distribution was that of the 1950s or the 2000s.)

When central air-conditioning was installed in the U.S. Capitol in 1928, it was still a novelty. Marsha Ackermann relates how, on a day in late May 1929, John Rankin, a Democratic representative from Mississippi, rose to complain that the thermostat had been set too low, that maintaining a difference of 15 to 20 degrees between the indoors and outdoors was "too much." He declared, "This is regular Republican atmosphere, and it is enough to kill anyone if it continues." The political debate over air-conditioning did not last long at all, and climate control has enjoyed eighty years of bipartisan consensus. But by playing an indirect role in shifting scores of seats in the House of Representatives to conservative Sun Belt states, air-conditioning contributed to the growth in the GOP's power in the Congress between 1981 and 2006. And despite the Republicans' dramatic descent from power in 2006 and 2008, the country's geography is as politically fractured as ever.

As a sunny, heavily Democratic state that grew rapidly through the age of air-conditioning, California would seem to be an exception to the trends just described. However, the state as a whole is no longer the magnet for domestic migration that it once was. For at least two decades, an average of 208,000 more domestic migrants per year have been moving out of the state than into it. Most of California's population growth during that time is attributable to immigration from outside the United States. And because the bulk of the state's population is crowded toward the coast and the chilly waters of the Pacific, the proportion of households with air-conditioning is smaller in California than in New England. In its overall politics and low aggregate space-cooling capacity, California's a blue state all the way. Within the state, however, political and climatic differences mirror those of the nation. As you look across from the cool coast to the often-hot inland lowlands, the more Republican the political climate becomes. The state's 2004 and 2008 electoral college maps show mostly blue counties on the coast and red counties in hotter portions of the state's interior.

In both electoral clout and climate, residents of red states hit a red-hot peak in 2004, experiencing 153 percent more cooling degree-days per year on average than did blue staters. Blue America warmed up a bit in the 2008 election when Florida, Nevada, Virginia, and North Carolina went Democratic in the presidential race. But the red states still saw 70 percent more CDD and 34 percent fewer heating degree-days than did the blue in 2008. Partly because of their lower level of urbanization and history of suburban sprawl, the red states of 2008 used 52 percent more total energy per capita (including 29 percent more for transportation) than did the blue states and emitted 123 percent more carbon per person. But neither the energy use nor the carbon emissions of red states should be cause for smugness in New York, Cambridge, or Berkeley. Just as China and India now take care of a lot of America's industrial "dirty work," the South and West these days get stuck with some of the nation's more heavily polluting industries.

Arsenault believes that, whatever its partisan political impact, air-conditioning has helped open up opportunities for cultural interaction in the South. "It's an interesting irony," he told me. "With air-conditioning, along with highways and air travel and mosquito control and other technologies, you got technological homogeneity in the South, which encouraged a greater variety of people to come in. So in contrast to the technological homogeneity, you got a lot more human variability." Arsenault compares the paradoxical result with that of the debate over bilingual education: "Opponents are opposed to multiculturalism, so they want 'English-only' laws in education. But ironically, that common language can allow greater communication among students and actually foster multiculturalism, the very thing that English-only was supposed to prevent. It's similar with air-conditioning, which has eroded traditional boundaries and created new possibilities."

But demographer William Frey, a senior fellow with the Brookings Institution, has seen new boundaries laid down in the age of air-conditioning. He wrote in 2002 that the cultural and political maps are now becoming more complicated as the urban-to-suburban "white flight" that helped shape the late twentieth century is replicated on a continental scale in the twenty-first century. Whites have begun moving in droves from what Frey calls "Melting Pot" regions—populous states with large populations of foreign immigrants, including California, Texas, Illinois, Florida, and New York—to the suburbs of a similarly discontinuous "New Sun Belt." Writes Frey, "In their exodus from the largely cosmopolitan, liberal-leaning urban areas, the participants in this new suburban flight are sharpening the differences—cultural and political as well as demographic—between the New Sunbelt and the Melting Pot regions." As whites become even more concentrated in homogeneous areas, Frey believes, greater social isolation will result: "What is missing from this new scenario is the opportunity that used to exist for daily, face-to-face interactions among people from these different social worlds."

Also in 2002, the Census Bureau predicted that the population of the Sun Belt (this time an expanded version that includes all the states of the West) would grow between 2010 and 2030 at almost seven times the rate of the population of the North, upper Midwest, and East—

what it termed the "Snow Belt." This expanded Sun Belt's advantage in the Electoral College, which stood at 88 votes in 2004, was predicted to swell to 146 votes by the 2030s. Despite including cool states like Colorado and Montana, the bureau's Sun Belt would still be 113 percent hotter than the Snow Belt in terms of population-weighted cooling degree-days. The bureau then announced in 2009 that economic stresses appeared to have caused migration to the Sun Belt—indeed, all movement of families around the nation—to slow dramatically. William Frey told the press that the report marked the "bursting of a 'migration bubble.' " Yet the country had already been shaped by several decades of migration; as late as 2007–08, ninety-four of the nation's hundred fastest-growing counties were in the South and West. And the recession may have represented no more than a temporary dip in migration volume. The Sun Belt does not look like it's ready to be written off just yet.

THE LIVING ROOM ON WHEELS

In 2005, American commuters were spending an average of two and a half times as many hours in traffic delays as they did in 1983. Eight of the ten cities with the worst traffic problems also experience more hot weather than all but one of the cities that have the least trouble with delays (see Table 4). Sun Belt cities with some of the nation's hottest summers are among those where drivers spent the most time sitting still or crawling in traffic, often with the air conditioner set to "max." Fifteen Sun Belt states stretching from Virginia to Nevada (the thirteen previously discussed, plus Tennessee, Arkansas, and Oklahoma) are home to one-third of the nation's motor vehicles and burn half of the fuel used to run vehicle air conditioners.

IBM's 2008 "Commuter Pain Index" is based on a poll of more than four thousand drivers in ten U.S. cities who were asked about time spent commuting, time spent stuck or in start-stop traffic, stress caused by traffic, and other variables. The goal, according to the company, was to "help IBM scientists as they develop and deploy new approaches to traffic mitigation like automated tolling, real-time traffic prediction, congestion charging, and intelligent route planning." Miami, Los An-

Table 4

With a few exceptions, cities with the worst traffic delay problems tend to have more hot weather, with a correspondingly higher air-conditioning requirement, expressed as EPA-estimated annual cooling hours per year. Included are metropolitan areas of more than one million in population. Traffic delay data are from 2007, as reported by the Texas Transportation Institute. (That year, New Orleans' traffic volume was depressed by a population loss of close to 40 percent in the aftermath of Hurricane Katrina.) Cooling hours are from EPA 2009.

City	Hours spent in traffic delays per traveler per year	Gallons of fuel wasted per traveler	Annual cooling hours
Most time wasted			
Los Angeles area	72	57	1,530
San Francisco–Oakland	60	47	224
Washington, D.C., area	60	43	1,320
Atlanta	60	44	1,484
Dallas–Fort Worth	58	40	1,926
Houston	56	42	2,209
San Diego	57	44	1,347
Orlando	54	35	2,915
Detroit	54	35	642
Miami	50	35	3,931
Least time wasted			
Providence	29	17	656
Cincinnati	27	19	996
Milwaukee	19	14	513
New Orleans	18	11	2,388
Kansas City	17	10	1,032
Pittsburgh	16	9	737
Cleveland	12	9	639
Buffalo	11	7	571

geles, Atlanta, and Dallas ranked at the top in driver misery, with pain index values from 4.8 to 9.6 on a 10-point scale. Those four cities experienced, on average, three times as many hours hot enough for air-conditioning as did the six cities further down the list: Washington,

New York, Chicago, Boston, Minneapolis, and San Francisco, with traffic pain ratings of 1.4 to 4.2. Over the ten cities, the average one-way commuting trip was 16.7 miles. IBM didn't ask its subjects whether heat added to their stress, but the commuter pain index in those big, hot Sun Belt cities would no doubt have been off the charts were it not for air-conditioning.

The earliest use of air-conditioning in transportation was in train compartments, and it spread rapidly through the rail system in the 1930s. Today, of course, trains handle a vanishingly thin slice of passenger transportation in most parts of the country. In getting to work daily, Americans are depending more heavily than ever on cars and light trucks. The national share of commuters using public transportation continued to fall during the 1990s, as did the share who carpool. Meanwhile, the amount of time that the average American wastes in traffic stoppages went from fourteen hours per year in 1982 to thirty-eight hours in 2005. By then, the resulting fuel wastage across 437 urban areas had reached 2.9 billion gallons per year.

The IBM survey found that drivers had plenty of ideas on how they'd rather spend the extra time they now pass in their vehicles. Forty-three percent would spend more time with family and friends; 34 percent would use the time for recreation; 31 percent would get more exercise; and 31 percent would get more sleep. Even though most Americans have no choice but to drive to work, is there a pain threshold beyond which we would no longer accept spending a substantial part of our waking lives strapped into a climate-controlled pod? Automotive commuting is one of the most widely tolerated annoyances in American life. It seems worth asking whether the working people of America would be in open revolt by now against the mind-numbing ordeal of ever-lengthening commutes were it not for air-conditioning, mobile phones, stereo systems, satellite positioning systems, lumbar support, and other conveniences, diversions, and comforts that have become an integral part of what we define as a car.

To improve the tolerability, if not the pleasure, of those hours on the road, automakers have gone to great lengths to create the perfect interior environment. Here is a review of the 2009 Lincoln Navigator from *U.S. News & World Report*: "Notable interior features standard on all trim levels include leather upholstery, 10-way power adjustable front seats with lumbar support, second- and third-row fold-flat seats, adjustable head restraints and first- and second-row floor consoles. Moreover, visor-imbedded vanity mirrors, a universal garage door opener, adjustable pedals, driver's seat/side mirror memory settings, a leather-wrapped steering wheel with audio/temperature/navigation controls and a dual-zone automatic climate control system with second- and third-row vents contribute to overall comfort."

That review somehow omitted the phrase "living room on wheels," which has been a staple of light-truck advertising since the mid-1990s. A quick Internet search shows that the Navigator—as well as Dodge Ram pickups and Grand Caravan vans, the Toyota Tacoma pickup, the Nissan Cube, the GMC Yukon, the Lexus RX450h, the Ford F-350 pickup, and just about every other gas guzzler on the market—has been plugged at one time or another as a "living room on wheels."

Early promoters of climate control stressed that air-conditioning is designed to help humans endure a "hostile environment." It's no wonder, then, that climate control is a requirement in the world of the commuter, who is under constant siege by unfriendly weather, foul air, accidents, construction delays, and road rage. In a 2005 paper, Josh Lauer, now an assistant professor of communications at the University of New Hampshire, echoed SUV ads of the time: "These oversized vehicles seek to recreate the comfort and privacy of one's living room inside the vehicle." But he dismissed the SUV's widely accepted reputation for safety and spaciousness, offering an alternative explanation:

> Safety is not road safety but personal safety, and space is not interior cargo space but social space, including the ability to traverse the most inhospitable terrain to sequester oneself from the hazards of modern civilization. In this way, the SUV's popularity reflects underlying American attitudes toward crime, random violence, and the importance of defended personal space.

In this analysis, air-conditioning can be thought of as part of a vehicle's armor, because it is often necessary even in slightly warm weather if the windows are to remain closed for safety's sake. If, to give the air

conditioner a small break, the windows are tinted, all the better; occupants need not fear appearing vulnerable to potential attackers. Lauer
adds, "More importantly, such one-way viewing affords those inside
the voyeuristic ability to surveil outsiders without their knowledge,
thus enhancing the power differential." And if American car buyers
drift back toward more fuel-efficient vehicle types as oil prices rise, it's
not likely that they will want to give up the illusion of security that
SUVs have done such a good job of creating.

FUEL CONSUMPTION FOR COOLING

Air conditioners that cool vehicles work on the same basic principles as
the residential unit in Figure 1, but the energy comes from liquid fuels.
About 5.5 percent of the fuel burned in a calendar year by America's
cars and light trucks—more than 7 billion gallons—goes to run their
air conditioners. Estimates using different assumptions put the total at
10.6 billion gallons. Mobile cooling has traveled fast across the globe.
In Europe, automobile air-conditioning was rare until the twenty-first
century. Now it's being adopted at a rapid clip, with more than two-
thirds of new cars air-equipped and 95 percent projected to have it
by 2020. The European Commission forecasts that the one-two punch
of increased fuel consumption and refrigerant leakage from cooling
systems will soon add 10 percent to greenhouse emissions from the
continent's vehicles. Adoption of air-conditioning is blamed for European car manufacturers' failure to meet voluntary greenhouse reduction targets for the 1998–2008 decade.

Mobile air-conditioning represents a large, durable addition to national resource demand. Warming a car's interior in winter consumes
little extra energy, because it uses waste heat from the engine. But air-
conditioning requires far more energy than any other auxiliary system
in a vehicle. With its air conditioner running full blast, the typical car
guzzles 19 to 22 percent more fuel per mile traveled than it does when
the air is turned off. That figure, from the National Renewable Energy
Laboratory, was based on a standardized driving pattern that includes
a combination of urban and highway travel. The tests did not include
creeping for miles in traffic jams, or the other types of forced idleness

common in urban driving; at those times, air-conditioning can account for the lion's share of the fuel burned. Cooling can cancel out much of the fuel-efficiency advantage for which hybrid electric vehicles are prized. One government study found that in urban driving, a hybrid can see its fuel consumption increase by 19 to 66 percent and a full-electric vehicle by 16 to 36 percent as the air conditioner control is moved from "Off" to "Low" to "High." Unless other measures are taken to reduce tailpipe emissions, an increase of x percent in fuel consumption results in an x percent increase in carbon dioxide emissions. And the EPA says that the global-warming potential of carbon dioxide produced by burning vehicle fuel for cooling is matched by the global-warming potential of refrigerants leaked during operation and servicing of mobile air-conditioning systems. The Coordinating Research Council, funded by the oil and automobile industries, estimates even larger impacts on emissions of other pollutants; in its tests, air-conditioning increased emissions of carbon monoxide and nitrogen oxides by 70 to 80 percent.

In the long-running debate over whether you'll use less gas on a highway trip by keeping the windows open (which increases the car's aerodynamic drag) or rolling them up and turning on the air (which puts an extra load on the engine), myriad tests of all kinds by the Society of Automotive Engineers, *Car and Driver* magazine, the *Mythbusters* TV show, and many, many others give widely varying results. Some support air-conditioning, others favor open windows, and few if any have much statistical heft. The debate should be declared a tie at this point; you can run the air-conditioning while cruising on an open interstate highway and be confident that you're not wasting much if any fuel. In slower nonhighway driving, with varying speeds, stops, and starts, air-conditioning always increases fuel consumption.

Vehicle air conditioners are cooling far more cubic feet of passenger space and running far more hours now than they did two decades ago. The total number of vehicle-miles traveled in the United States has doubled just since 1990, rising more than four times as fast as the population. Americans finally cut back on vehicle travel starting in 2008, but most who still had jobs still had to commute, and we continued to do so in mostly empty vehicles. In more than 85 percent of commuting vehicles, the driver rides alone. Despite being prized for their roominess, light trucks, a category including SUVs, vans, and pickups, haul only 1.7 people each, including the driver, on an average trip.

In 2009, the California Air Resources Board decreed that by 2012, windshields and windows on all new cars sold in the state must incorporate tiny metal flakes that deflect solar radiation. The windows will be required to reduce cars' absorption of energy from sunlight by 45 percent in 2012 and by 60 percent in 2016. The board estimated that the measure would cut carbon dioxide emissions by 700,000 metric tons in 2020; perhaps more important, it would reduce levels of deadly air pollutants in Los Angeles and other cities. A reeling auto industry whined about the increased new-car costs that would result. But for the driver, the costs could be made up in reduced gas consumption—within four years for the 2012 windows and twelve years for the much more expensive 2016 version. An earlier rumor, spread by radio host Rush Limbaugh and others, had warned that the board also planned to ban vehicles with black body paint. That proved to be unfounded.

An astonishing 5 to 8 percent of an average passenger car's fuel consumption takes place when it's parked or stopped in traffic, idling. Because miles per gallon have dropped to zero at that point, the fuel being burned is entirely wasted unless it's being used to perform some other work. The most common "other work" is to run the air-conditioning or heating system. In searching for "low-hanging fruit" (simple ways in which energy use could be cut painlessly and quickly), three Vanderbilt University professors argued in 2008 that the wasted fuel consumption of idling vehicles could be cut dramatically if more Americans understood that modern cars don't need a long warm-up period before being driven, and that cars use more gas while idling, even for a few seconds, than they would in stopping and restarting. They estimate that cutting idling by only 10 percent nationwide would save 570 to 900 million gallons of gas each year and keep 6 to 9 million tons of carbon dioxide out of the atmosphere.

In choosing as their conservation target a very modest 10 percent voluntary reduction in idling, the Vanderbilt team was being timid but realistic. Even if all Americans were fully versed in automotive fuel-consumption patterns and the evils of idling, the practice would continue. Without an outright ban on idling, we will continue to see cars, pickups, and SUVs scattered by the millions at curbsides and in parking lots and drive-through lanes, their engines running, all for one overriding purpose: climate control. The price of gas can rise and fall and rise again, but when the weather is hot, you will see drive-up lines lengthen at banks and fast-food restaurants, their customers cowed by the prospect of a short, exposed sprint across a blazing-hot parking lot. You'll see vehicles left running in big-box parking lots, sitting locked and empty, their engines devoting 100 percent of fuel consumption to keeping the interior cooled in anticipation of their owners' return. Or the air conditioner may be running protectively, perhaps to keep a bag of groceries or a Shih Tzu puppy from turning to mush.

When fuel wastage by idling engines is discussed, drive-up windows at banks, restaurants, and other establishments are commonly fingered as top offenders. Consider the string of fast-food drive-throughs along a single five-mile stretch of Range Line Road in Joplin, Missouri. A local journalist calculated that those drive-throughs cause one hundred thousand gallons of gas to be wasted annually.

Banning drive-ups and giving tickets for idling would be relatively simple ways to curb fuel waste. But as long as the personal car or truck is our primary means of transportation, fuel consumption and pollution cannot be cut as deeply as they must be cut. If we do manage to reduce both miles driven and emissions per mile while at the same time clearing ample room in cities for nonmotorized travel and ensuring that mass transportation is as available, convenient, and comfortable as possible, life will be a lot more pleasant all around. Commuters in particular will be much less likely to get hot under the collar.

5

THE BUSINESS CLIMATE

General Electric has proved a more devastating invader than General Sherman.

—Raymond Arsenault, "The End of the Long, Hot Summer," 1984

A May 2003 survey by the International Facilities Management Association in Houston says being too cold was the No. 1 office complaint, followed by being too hot.

—New York Times, January 23, 2005

In the early days of air-conditioning in America, the urban rich tended not to show much interest in the novelty of mechanical cooling. The masses had their air-conditioning, the wealthy their beach houses. Marsha Ackermann explained that upper-class Americans "had an inbred habit of ignoring discomfort." Their tolerance of heavy warmth, it was believed, extended to the retail realm: "The slow acceptance of storewide air-conditioning reflected not only its high cost but also a common belief that gentlewomen, and the type of women fit to serve them, were by nature and culture less bothered by heat." Indeed, air-conditioning was first installed in the "bargain basements" of department stores, not the upper floors where luxury goods were on offer.

The belief in a link between upper-class status and tolerance of harsh conditions was not easily reconciled with another idea (one friendlier to the emerging air-conditioning industry) being put forth by some academics of the time: that mild temperatures favor economic development and prosperity. The two ideas were not mutually exclusive, but harmonizing them depended on a new amalgam of genetic and social theories. Geographer Ellsworth Huntington, called "the high priest of climatic determinism" by historian A. Cash Koeniger, amassed mountains of data, processed them through sometimes confusing logic, and claimed to demonstrate that people living in too-warm climates were incapable of building and sustaining prosperous economies. He and other climatic determinists popularized the idea that heat inflicts physical and mental weakness on entire populations. But was the imagined weakness "caused" by the heat or did it indicate genetic vulnerability to heat? And, once weakened by heat, could people be reinvigorated by a cooler climate, either natural or artificial?

ENERGIZING THE SOUTH

With the publication of his book *Climate and the Energy of Nations* in 1944, Briton Sydney F. Markham attempted to clarify matters by combining the concepts of outdoor and indoor climate into a single formula that did not depend heavily on shady human eugenic theories. Markham saw an almost perfect correspondence right across the globe, at large and small scales and in all eras, between moderate temperatures and a high "energy" level in societies. "Energy" in his title referred not to oil or coal but to the physical and mental energy that humans expend in building politically and economically powerful nations. Genetics had nothing to do with it, he insisted; formerly successful populations invariably slid downhill within a generation after migrating to a too-hot climate. Where heating technology had tamed the winter in cooler regions, he concluded, economic and social development had reached its peak because people were experiencing temperatures nearest to what he had determined to be the ideal year-round average: 70°. Heating technology had allowed the northern United States to gain superiority over the once-dominant South in the 1800s. Southerners, meanwhile, having lived through many long, hot summers, had adapted to levels of heat that other Americans found oppressive, according to Markham. But their perfectly sensible adaptation to heat had held them back economically:

> The inhabitant of Kansas, Missouri, or Indiana normally works at a pace that is too fast for his hot summer. The native of Louisiana, Texas, or Florida, on the other hand, has acquired a lifelong habit of acting more slowly than the Northerners, so that

he does not suffer so much from the heat and perhaps would say that he enjoys it. The central point is that man adjusts himself to a rate of activity appropriate to the combined effects of the natural climate in which he lives and the artificial climate which he creates. In this connection, two things are perhaps ominous for the United States. One is that no less than 22 states out of 48 have from 1 to 5 months of the kind of weather that almost inevitably will continue to make people slow until some means of correcting it is found. Perhaps air-conditioning will do the trick, but thus far the mitigation of the effect of heat has been much more difficult than that of cold. Moreover, we cannot change the outdoor temperature, and people are bound to be out of doors a great deal in hot weather.

Markham put his conclusions in the form of a map in which he assigned each state a shading pattern based on its level of "civilization"— a variable within which he included rankings for infant mortality, per capita income, and "intelligence." The entire Sun Belt was ranked as having the lowest level of "civilization," as were frigid North Dakota and Maine. Having adapted too well to their hot climate, the people of the South were failing to achieve economic growth, asserted Markham, and he was not optimistic about the potential of air-conditioning to invigorate the region. Because "its influence ceases on the doorstep," there was "no indication yet that air-conditioning will help the rural South. . . . Until air-conditioned jeeps and tractors are made, the southern rural workers are likely to remain much as they are to-day." With his focus on physical outdoor work, Markham did not foresee the forces that, aided by air-conditioning, would transform the South, starting within a decade of his book's publication: industrialization, urbanization, migration, growth of suburbs, the struggle against racial oppression, and, eventually, the predominance of white-collar work. Air-conditioned tractors did indeed appear on the scene, but by that time, the more sweeping transformation was already well under way.

As air-conditioning became common across America and more work moved indoors, academics shelved climatic determinism alongside their other outdated theories. But in the late 1980s, Koeniger dared once more to delve into the relationship between climate and Southern culture. In doing so, he chided fellow scholars who sat in air-conditioned offices and waved away any mention of climate as a historical force. He asked, "[I]s it merely a coincidence that the first generation in the history of mankind freed by technology from virtually all unwanted contact with climatic environment was the first also to dismiss the importance of climate?"

By moderating that part of the environment made by humans, air-conditioning stimulated the economic rise of the South, first in manufacturing and then in white-collar industries. As early as 1920, air-conditioning was being depended upon to control temperatures—and, more important, humidity—in a number of the South's signature industries: cotton, rayon, tobacco, paper, and baking. But the foundations of those industries, and of the economy in general, were still in agriculture. That agriculturally rooted economy has not disappeared; consider the South's domination of poultry processing, which could not be accomplished on its current gargantuan scale without big outlays for refrigeration. But the nonagricultural economy has far outgrown the traditional industries. Arsenault writes, "After decades of false starts and inflated promises, industry came to the South in a rush after World War II. . . . Some commercial and industrial growth would have occurred in the post–World War II South without air-conditioning. But the magnitude and scope would have been much smaller."

In 1983, Mancur Olson of the University of Maryland attributed the South's rapid postwar growth almost solely to the availability of cheap labor. But why did that growth start so late? After all, the Old South had had a large, poor population of exploitable workers since the nineteenth century. Olson recognized that the region's hot environment, as long as it remained untamed by air-conditioning, was a "significant factor" in delaying economic growth for so long, but unlike Sydney Markham decades earlier, Olson determined that climate was "only part of a much larger story." He pointed out that cold weather is also unpleasant and raises production costs as much as do heat and humidity; that postwar growth was rapid in some Northwest and Plains states as well as in the South; and that, historically, economies have also grown rapidly in hot, non-air-conditioned places like

Mesopotamia, the Nile and Indus valleys, and the Mediterranean. He also predicted that with their then-maturing economies and rising wage rates, the states of the South would see their economic growth come down to the same rates as those in the rest of the nation. Concluded Olson, "The South will fall again." He didn't specify when that would happen, and for another quarter-century the South kept outperforming the North economically. In the years before 2008, the Sun Belt even widened the growth gap, as can be seen in Figure 2.

Olson's prophecy finally came to pass in 2008, with the rest of the country joining in that economic decline. The thoroughgoing changes

Figure 2

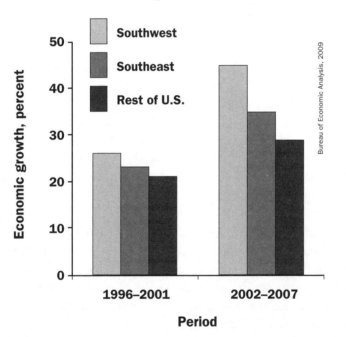

Total economic growth (expressed as percentage of initial gross regional product for each interval of five years shown) has been greater in the South and Southwest than in the rest of the country in recent years, and the gap was widening even as the economic downturn began. These percentages were calculated from U.S. Bureau of Economic Analysis (BEA) figures for state gross economic products. Regions are defined by BEA.

that swept over the South in earlier decades will, however, endure. Here we find yet another economic spiral: air-conditioning helped create a more attractive climate for manufacturers and employees in business-friendly Southern states, and the freedom to pay lower wages spurred profitability and growth, which attracted further migration to the South. Without air-conditioning, the Raleigh–Durham–Chapel Hill triangle in North Carolina would never have grown to become a powerhouse of biotechnology, pharmaceuticals, and computing research. The automobile manufacturing plants that now litter the South probably would still be back up north or in Japan and Korea. Jacksonville would not have become an insurance and banking center, and Birmingham would not have moved from steel into health care and communications. The world's first and seventh busiest airports would not be located in Atlanta and Dallas. And without air-conditioning, of course, the first domed stadium could not have been constructed in Houston in 1965, and the country's five largest indoor stadiums today would be sited in the frozen north rather than in Arlington, Texas; Houston; Glendale, Arizona; New Orleans; and Atlanta.

BRISK SALES

The ghost of Sydney Markham has never fully faded from the economics literature, because the relationship between nations' annual average temperatures and their average levels of income and growth appears to be so strong and so obvious. Although comparisons between rich temperate countries and poor tropical ones show a negative relationship between average temperature and economic activity, comparisons of regions within countries tend to show the opposite—that warmer temperatures actually boost economic activity. It is not easy to separate the effects of climate from those of history, including the economic and military exploitation of poor countries by rich ones. A recent comparison *within* countries between cool and warm *years* attempted to separate those forces statistically. It showed that economic growth is stronger under lower temperatures—but only within poor countries. Rich countries' economies did not appear susceptible to temperature fluctuations. The authors of the study wrote,

We also find evidence for a broad set of mechanisms through which temperature might affect growth in poor countries. While agricultural output contractions are part of the story, we also find adverse effects of hot years on industrial output and aggregate investment. Further, higher temperatures lead to political instability in poor countries, as evidenced by irregular changes in the national leadership. These industry, investment, and institutional effects sit outside the primarily agricultural focus of most economic research on climate change and underscore the importance of an inclusive approach to understanding climate change implications.

Thanks to technological buffers—air-conditioning prominent among them—wealthy industrial nations typically manage to avoid the economically depressing effects of hot weather. Since before World War II, American businesses have been built around the idea, right or wrong, that air-conditioning can boost the productivity of workers and also stimulate customers to buy. In the age of air-conditioning, the mass-retail industry would never go back to the summertime doldrums of the 1920s, as described by Ackermann: "[S]ummer was when saleswomen's feet swelled, customers stayed away, and the perspiration of the few who shopped turned stylish crepe and georgette dresses into 'drab, crushed' markdowns." In describing the "Hot America" of those days, a turn-of-the-millennium exhibit at the National Building Museum in Washington, D.C., painted a more benign picture, one of a nation with sagging summer productivity but also with better things to do than to go shopping. It read in part,

> Before air-conditioning, American life followed seasonal cycles determined by weather. Workers' productivity declined in direct proportion to the heat and humidity outside—on the hottest days employees left work early and businesses shut their doors. Stores and theaters also closed down, unable to comfortably accommodate large groups of people in stifling interiors. Cities emptied in summers. . . . Houses and office buildings were designed to enhance natural cooling, and people spent summer days and evenings on porches or fire escapes. They

cooled off by getting wet—opening up fire hydrants, going to the beach or diving into swimming holes.

Gail Cooper wrote about industry executives arguing forcefully in the 1930s that "air-conditioning in particular could be a powerful force for stimulating the economy and ending the Depression." Not only would it provide manufacturing jobs, it would also help electric utilities by soaking up some of their spare capacity during times of reduced summertime lighting demand and bring brisker trade to the retail sector. Its economic importance was implicitly recognized by the federal government in 1942, when the War Production Board banned the manufacture or installation of air-conditioning systems "solely for personal comfort." Systems were removed from a few big-city retail stores and installed in military production facilities. The program was terminated by 1944, and air-conditioning went on to become a pillar of the postwar economy. The commercial world has embraced air-conditioning more strongly than ever in recent times. In just eight years, from 1995 to 2003, energy expended to cool the retail sector shot up by 66 percent. About half of that increase resulted from more intense cooling per square foot.

Back when society followed "seasonal cycles determined by the weather," keeping consumer demand revved up was no easy task. Now, things are different. In the service economy, one of air-conditioning's crucial functions is to eliminate perspiration, the human body's key defense against heat stress. Public sweating in indoor spaces other than gyms and health clubs is now viewed as a breach of etiquette in most urban and suburban areas. The usual rule today is that if sweating can be avoided, it should be avoided. Interviewing young residents of Singapore, scholars Russell Hitchings and Shu Jun Lee learned that air-conditioned commercial spaces are popular in the tropical city because they make Western-style clothing, especially highly popular black outfits, more comfortable. The chief concern of the interview subjects, all of them in their twenties, was sweat: "Bodily wetness in the form of perspiration emerged as an especially detestable condition, except when exercising."

The common view of sweat as a symptom of actual illness appeared to be encouraged by the Singapore government. According to Hitchings and Lee, "In 2001, [Singapore's] environment minister paid tribute to air-conditioning, which had seemingly enhanced productivity to the extent that this technology was 'one of the reasons why Singapore now enjoys the highest per capita GDP among countries on the tropical belt.'" He had even identified "the humble air conditioner" as "the millennium's most influential invention." The young people whom Hitchings and Lee interviewed had, they concluded, "internalized a version of the human productivity argument espoused by the government in terms of the importance of fixing indoor ambiance at specific human standards since, to do otherwise, would mean that people simply could not function effectively."

The days when the upper classes supposedly strode briskly through the summer heat, leaving artificial cooling to the rest of us, are long gone. On a humid 79° day in June 2005, the *New York Times* confirmed what many suspected: the more upscale retail establishments of Manhattan push their air-conditioning systems harder. "In other words," wrote reporter Allen Salkin, "the higher the prices, the lower the temperatures. Consider these clothing stores: Bergdorf Goodman, 68.3 degrees; Bloomingdale's, 70.8; Macy's, 73.1; Club Monaco, 74.0; the Original Levi's store, 76.8; Old Navy, 80.3." A Bergdorf Goodman vice president defended the store's chilly conditions: "It's part of the whole environment package that we try to offer our customers. We're offering the best of service in New York City, and what comes with that is how the store looks, how it's lit, the cleanliness, and the temperature." A retail consultant told Salkin, "There is still status symbol in almost over-the-top air-conditioning."

Whatever the prestige level or thermostat setting in the store that's doing the selling, home and vehicle air-conditioning can make major purchases more attractive. Full enjoyment of a jumbo-screen TV, a PS3, a DVD, a PC, an SUV, or an RV calls for A/C. Air-conditioning allows you to grill steaks in a comfortable kitchen, hit simulated golf drives when it's too hot outdoors, or, as President Richard Nixon used to do, enjoy a real fire in the fireplace any time of the year.

There is more retail space to cool than ever. The expanse of shopping territory available to each American continued to grow right up to the 2008 recession, and the volume of air being cooled in the merchandising world reached record levels. Per shopper, the amount of mall space (including enclosed and strip malls) grew by 34 percent in just eight years. Retail uses 16 percent more electricity per square foot for air-conditioning than do offices. It consumes more total cooling energy than do offices or educational institutions, and three times as much as do health care facilities. Any of those big-box and mall spaces left standing empty by the recession represent a lot of lost income and material waste, but their closing also means a huge savings of energy if they can manage to stay empty.

EXTREME INDOOR CLIMATES

Manufacturing jobs migrated southward within the country in the age of air-conditioning and even more swiftly out of the country since 1990. But 10 percent of the U.S. workforce is still employed in manufacturing. People who work in factories often endure greater ranges of temperature and humidity than do workers in the service sector. The often extreme working environments and sheer size of many factories make comfort-cooling exceedingly expensive, and a company's average wage investment per worker usually doesn't justify it. There is also a common belief that the higher the ratio of physical to mental work, the smaller the effect of too-hot or too-cold conditions. But all work involves some degree of mental performance, and the brain doesn't fare well in extreme heat. Common effects of extreme heat on workers' performance include inconsistency, inability to concentrate, negativity, drowsiness, headache, fatigue, and vulnerability to accidents. In heavy-industrial settings, "unsafe work behaviors" follow a U-shaped curve in relation to workplace temperature, with the safest temperature range usually found to be in the high 60s or low 70s. Unsafe practices increase as the air gets cooler or warmer than that. A study of workers in a metal-products plant and a foundry found that unsafe behavior among employees under moderate workloads was 85 percent more common at 95° than at 70°. The U-shaped response, noted the authors of the study, was consistent with the way human performance responds in general to "increasing levels of environmental stimulation or arousal."

In setting our home thermostats, we are free to decide how much money we're willing to spend for comfort. But in a factory, no individual has that freedom: the bottom line dictates the temperature. The story of a Columbia, South Carolina, plant owned by Plasti-Line, Inc. (the country's largest "single-source supplier of exterior image and signage products," today known as ImagePoint) is instructive. The company opened a new 82,500-square-foot plant in 1998. That first summer, workers throughout the building endured indoor air temperatures that could climb as high as 110° and stay there. A trade magazine's story on the plant, where plastic sheets are baked in open ovens at up to 350°, noted that managers were eager to cool the building down a bit because they knew that "deliberate work slowdowns, walkouts and similar job actions occur over heat problems more than any other workplace hazard." But a minute or two with a calculator told them that installing and operating an air-conditioning system would shatter their profits.

They also considered and rejected seemingly sensible advice from the federal Occupational Safety and Health Administration: to assign lighter work to employees who needed time for heat adaptation and to allow longer rest periods in cooler areas. Managers calculated that a single daily rest period of ten minutes for their hundred-person workforce would cost them $20,000 over a summer—only 2 percent of their payroll over that period, but still a lot of money. Managers settled on a remedy that would cost only about $600 per summer in electricity bills and would not slow production: installation of nine high-tech hollow-core aluminum ceiling fans, each twenty feet in diameter. Each of the giant, slow-moving fans (made by a Lexington, Kentucky, company now called Big Ass Fans, Inc.) could bathe 15,000 to 20,000 square feet in a three-mile-per-hour breeze, which would be enough to evaporate perspiration efficiently and cool the body significantly without blowing the ovens' flames out. The air temperature was not lowered, but employees were better able to shed heat from their bodies.

The fan array cost one-twenty-eighth as much as air-conditioning to install and one-tenth as much in operating power. A manager, as quoted in the article, concluded that "the air flow provided to the employees has been a huge relief." The story included no quotes from employees, who might well have preferred to have both the improved air circulation and more frequent breaks from the heat. In simulations across a range of climates, Big Ass Fans analysts have estimated that maintaining air movement at 1.8 miles per hour (a speed limit imposed by ASHRAE for office cooling, because stronger breezes blow papers off desks) saves 7 to 10 percent over warm-weather energy consumption in less extreme, already air-conditioned spaces.

In many plants generating less heat, the opposite approach can keep an occupied working space cooler. Thermal stratification is a way of using the hot air that collects just below a ceiling to insulate the cooled space below and transfer heat out of that space at night. It requires that the ceiling be relatively high and that the upper layer of indoor air be kept very still. Consider a North Kansas City, Missouri, industrial space once used by an air-conditioning manufacturer. With a twenty-foot-high ceiling and flat metal roof, the thirty-thousand-square-foot building had an air-conditioning system only about 60 percent as powerful as is normally used in such a space. The efficiency lay in the placement of the outlets for cooled air and return inlets for warm air. They hung down well below the ceiling, with inlets and outlets about ten feet above an exposed concrete floor and the outlets directed downward. In such a design, the upper layer of air can become very hot, and as long as it remains undisturbed, little heat moves into the occupied space. At night, any heat that the concrete floor had absorbed during the day was radiated efficiently to the ceiling, and the roof, having been heated to a very high temperature by the hot air and ceiling below, radiated heat to the night sky. Even at times when the temperature of the layer of outside air just above the roof reached 110° or more, and air just under the ceiling was 94°, the occupants walking across the concrete floor felt temperatures varying only between 68° and 74°, and that was accomplished with very low energy consumption for air-conditioning.

When the workplace does have to be air-conditioned for the sake of machinery or materials, workers may become overchilled. In few places is this as obvious as in the meat-handling industries. In its 2005 exposé of worker abuse in poultry processing, Human Rights Watch included portions of a court transcript from a lawsuit against Tyson Foods, Inc. In it, a plant manager testified about the company's decision to install "massive air conditioners" in one of its Virginia plants to lower the working temperature from the high 60s to 50°. Corporate calculations had shown that the resulting high electricity bill could easily be compensated for by reduction or elimination of midshift sterilizing "washdown time." Bacteria multiply much more slowly at lower temperatures, so the intense air-conditioning allowed production to continue without being interrupted by sterilization crews. Apparently, the economic returns of chilling the work space outweighed the increased electrical bills. The price for that extra output was paid by plant workers, especially older workers. The manager testified that the 50° environment was "so hard on them, they were complaining of bursitis, arthritis, and increased musculoskeletal problems." Some simply had to quit their jobs.

TUG-OF-WAR AT THE THERMOSTAT

As the majority of jobs have moved from the factory to the office, hospital, or retail store, the worker's thermal environment has come under greater scrutiny. It is widely believed that in the white-collar world, comfort and performance are more tightly linked than they are on the factory floor. A Wall Street stock trader told the *New York Times* that the room where he worked was kept "icy" because "we get stirred up during big trades, and we'll complain if it's too hot." He added that, during slack periods, "we all have fleeces we wear."

For decades, conventional wisdom has said that people's minds respond much like computer chips, working faster and more efficiently as the thermometer drops. As a result, air-conditioning systems have become almost universal in the business world. The trend's origins go back to 1902, when the first modern comfort air-conditioning system was installed, aptly enough, at the New York Stock Exchange, but office air-conditioning remained rare for decades after that. Today, as a result of employee demands, ergonomic experts' advice, and architectural fiat, the American office is, by definition, a refrigerated workplace.

In the big move to refrigerated climate control during the 1950s and 1960s, office buildings followed trends first established in the residential sector. As we've seen, "contagion diffusion" was not responsible for air-conditioning's popularity in homes, but for business firms, it mattered very much whether or not their neighbors had installed it.

By the reckoning of top officials at the Carrier Corporation—the pioneer of large-scale climate control equipment—once 20 percent of office buildings in a city were air-conditioned, competitive pressures compelled the rest to follow. Office buildings were reconceived as massive cubes; because they were expensive to build and air-condition, window-accommodating H-, T-, and L-shaped footprints were out. Air-conditioning constituted 16 percent of the cost of a 1950s-era office building, the second-biggest expense after steel. Builders bore the cost nevertheless, according to Gail Cooper, in order to maintain the "appearance and popularity of the modern office block," which was being constructed more and more often with the "lavish use of glass." With electricity and mechanics substituting for windows, the "deep space" within the block—square footage that cost the least to build— could be made comfortable and profitable. Unlike houses, office blocks needed air-conditioning even in cool climates. I found it depressing to read in Cooper's book the words written in 1932 by a Carrier publicist as he looked forward to the day when air-conditioning would "enable the prescient architect to go about his designing unfettered by the erstwhile necessity of 'ventilating shafts,' 'light wells,' 'outside exposures,' and such considerations." All those features, which the publicist put within dismissive quotation marks, are today proposed as staples of the "green" architecture of the future.

As it turned out, building the climate-controlled workplace was the easy part; deciding upon the right indoor climate—one that would make all employees happy and productive—was much more difficult. Still today, on the seemingly simple question of where to set the thermostat, neither research nor practical experience has provided firm answers. Struggles over office temperature arise because human bodies do not come with instructions that include a recommended range of operating temperatures. Studies of office comfort occupy acres of slick paper in academic journals, and their conclusions vary widely. But considering the stakes, the specific relationship between workplace temperature and employees' *productivity,* whatever their degree of personal comfort, has drawn the attention of surprisingly few researchers. David Wyon, professor at the Technical University of Denmark, is one of the exceptions, having spent forty-seven years studying the effects of the indoor environment and comfort on productivity. Although he says, "There is still a great deal to be done," he is confident in his basic conclusions: "Our studies show that human performance improves substantially at temperatures at the lower end of the comfort zone [around 68° to 72°] and at ventilation rates that maintain carbon dioxide levels well below 1,000 parts per million," compared with the current outdoor level at 350 to 400 parts per million and rising. "The size of the effect is more than enough to pay for the additional cost of achieving these conditions." Wyon believes that one mechanism through which low temperatures improve performance is "the accompanying decrease in building-related symptoms, both specific (for example, on the eyes) and general (for example, headache, difficulty in concentration, and fatigue), but that there are some direct effects on performance as well, accompanied by observable changes in the level of physiological arousal and motivation to exert effort."

There's a lot riding on researchers' ability to find the right temperature, if Wyon and his colleagues are right. He has written, "If it can be shown that the environment is among the factors that affect productivity, it becomes part of the cost-benefit calculation In the last analysis, [a nation's] economy, and thus society itself, is dependent on this calculation." He estimates that "improving the thermal environment in U.S. office buildings would result in a direct increase in productivity of 0.5% to 5%, worth $12 billion to $125 billion annually."

Employers who are not careful about temperature control risk an attack of "sick building syndrome." With routine buildup of pollutants in the interior atmosphere, virtually all buildings become at least a bit unwell. Accordingly, as much as a 5 to 13 percent productivity loss is thought to be "built into" tight, energy-efficient buildings, even when they adhere to international construction codes. Here, another feedback loop between pollution and energy consumption is evident. Wyon and co-authors have shown that warm conditions not only harm concentration, creative thinking, and general performance, but they can also amplify irritation by indoor pollutants. One trial, conducted in a real office but under controlled conditions, featured strips of twenty-year-old carpet from a reputedly "sick" building that were hung out of sight behind a partition "to ensure a moderate level of indoor air pollution" of the kind that caused employee complaints in the original building. Sets of employees were given simulated office work to perform over a series of five-hour periods and were exposed to various levels of temperature and humidity in randomly assigned blocks of time. Resulting graphs showed a straight-line relationship between air-conditioning intensity and worker satisfaction (i.e., lack of pollutant-related complaints): As temperature readings dropped from 79° to 68° and humidity was reduced, satisfaction with the environment steadily increased. The results had a lot to do with that nasty carpet behind the screen. Wyon and his co-authors surmised that "cooling of mucous membranes probably caused the air to be perceived as fresher and therefore improved the acceptability of air at low temperature and humidity." With higher levels of cooling, outside ventilation—the usual remedy prescribed for poor indoor air quality—could be cut back, putting a smaller load on the air-conditioning system. Despite its beneficial effects on perceived air quality and worker satisfaction, cooling had no effect on work output in that study.

I asked Wyon whether, by masking sick-building problems, intense cooling of recirculated air could allow tighter construction, reduced input of warmer outdoor air, and thus achieve a net reduction in energy use. He responded, "That is a purely engineering problem. In many climatic regions it is clearly possible, while in the hotter and more humid regions of the world it is more of a challenge." And would employees' well-being be sacrificed under that strategy? Wyon believes it would: "Eliminating indoor sources of pollution can greatly reduce the ventilation requirement, but cooling, while beneficial in itself, is not the right way to combat poor indoor air quality. It affects perception of air quality without removing the negative effects of poor quality. We are examining the possibility of continuous removal of air pollutants as an alternative to dilution with outside air." Removing pollutants will require filters of a kind yet to be developed; Wyon warns that indoor air filters currently on the market have "major problems."

Cornell University professor Alan Hedge rejects the widely held view that cool air makes people feel better and stimulates them to work harder. His research has led to the "heretical" conclusion that the way to elicit better productivity from employees is to warm them up. He told me, "Up until now it has been widely assumed that cooler temperatures promote better performance, but our results suggest the picture is more complex." Hedge and his colleagues have used unobtrusive keystroke-logging devices to measure output and error rates in studies at a Florida insurance company in winter, at the sales and accounting offices of a New Jersey factory in spring, and at two offices of a Long Island law firm in fall. In Florida, for example, when the thermostat was raised from 69° to 76°, keying output rose an astonishing 150 percent, and errors dropped by 44 percent. That, Hedge and colleagues reckoned, could boost profits by $2 per worker per hour.

Some studies have shown, Hedge concedes, that when people are made slightly uncomfortable by a chilly work space, they perform a little better. "But those are usually short-term temperature studies, up to three hours at a time," he says, and the effect is reversed when people are exposed to overly cool air day after day. "Think about it. Your core body temperature is 98.6°. Your skin is normally in the low 90s. When you walk from the 90° outdoors into the high 60s, it may feel great at first, but after a couple of hours, your body doesn't like it." I asked Hedge about the fact that none of his studies were done in summer. What difference might that make? He pointed to the adaptive model of comfort, which says that "you adjust your comfort level to the weather of the past month or so. That means the body might sense a too-cold office as being even more uncomfortably cold in summer." Hedge's work has drawn a lot of attention because it suggests that employers can cash in twice, by saving on air-conditioning bills and getting more work out of employees at the same time. (They might also be tempted to dial the furnace up to higher temperatures in wintertime to ramp up employee output, but that would mean spending money to make money.)

David Wyon cautions that the strength of Hedge's study—that its subjects were doing their real jobs, not experimental exercises—was also its biggest limitation. At any point in the day, employees were free to choose from among a variety of actions, including keyboard entry, studying the computer screen, and using the mouse. He thinks that as employees really start feeling warm, they spend less time on mousing, searching, and reading and more time on the mentally less-taxing work of keyboard entry. That, he says, would have produced the result seen in Hedge's data: high hourly keystroke rates at higher temperatures. Hedge says he has discussed these issues with Wyon, and, naturally, he disagrees. Mouse clicks did not decrease with temperature, and, he says, "the jobs we analyzed were either insurance or sales data entry, so there's not much discretion for higher cognitive tasks." Most of Hedge's subjects were women, and many of them "told the researchers that their offices were generally too cold, so maybe there is a gender effect." But there were too few men in the studies to test that possibility statistically.

Whatever indoor temperatures best serve the interests of business owners and managers, it is unusual for employees ever to reach unanimity on the most comfortable office temperature. Reporting on Hedge and his colleagues' work, the *New York Times* sought out a few battlegrounds in the office "thermostat wars." Typical was a skirmish between Adam Korn and Andrew Keown, office mates working for Hilton Hotels in Beverly Hills, California. Keown started work at the company in the heat of August and was immediately warned by other employees that Korn kept the thermostat in his area of the office at a very cold setting. Keown's predecessor, a woman, had worn a sweater in summer. Keown found the warnings to be accurate and "devised strategies for staying warm, like spending an extra few minutes in the bathroom and driving around after work with no air-conditioning in 80-degree heat." After a while, he also took to wearing sweaters. Finally, an increasingly desperate Keown dared to turn up the thermostat, "hoping Mr. Korn wouldn't notice." Korn did notice, and after some tit-for-tat adjustments, the two men reached a compromise temperature, one that moved Korn to keep a fan on his desk. Keown apparently considered but rejected a tactic that, according to the *Times*, is becoming common summertime practice in air-conditioned offices: keeping a space heater under his desk.

Research shows that women tend to complain about cold office conditions most consistently. Hedge says there is good reason for that, and it has nothing to do with biological differences between the sexes: "In the real world, the major difference is clothing. Men dress more heavily, and especially when it's cool, they will keep their ties and jackets on." And where we sit can have as large an effect on comfort as does how we dress. The Government Accountability Office found that as recently as the year 2000, female employees disproportionately occupy lower-level, nonmanagement jobs in retail, finance, insurance, real estate, education, and medicine. Gender distortion at the very top can be even greater; in balmy Florida, a 2009 study found that 92 percent of the executive positions in the top 150 companies were held by men. One result, says Hedge, is that men disproportionately experience the excellent insulating qualities of the typical "executive chair." "It's easy to imagine the boss sitting in his big chair in coat and tie giving the order to turn up the air-conditioning," says Hedge, "while his assistant is sitting in the next room, in a flimsy clerical chair in light clothes, already shivering."

U.S. employers are likely to continue overcooling office buildings, say analysts, even in the face of evidence that it does not raise productivity. Their real concerns may lie less with employees' comfort or performance than with company image. Unlike factories, many white-collar work spaces play host to a steady stream of outsiders. Managers want to keep visitors comfortable, but they also want to avoid presenting the image of a company whose offices feature an atmosphere even faintly reminiscent of a locker room. It is well established that the people stop noticing many odors after continuous exposure for about an hour, so smells that would have little effect on employees would be noticeable to arriving visitors. Like the low temperatures with which Bergdorf Goodman beckons its upscale retail customers, the heavy air-conditioning of offices is part of a "whole environment package" that signals a seriousness about doing business.

Struggles over office comfort may never be fully resolved, and the use of indoor climate control to improve productivity remains a tentative science. Even if the right formula is found, office workers may stand to gain little economically from their increased output. Productivity per employee rose dramatically over the past half-century as nonmanagerial wages stagnated and greenhouse emissions soared. The total number of working and commuting hours per household per week has pushed upward just to maintain, not increase, living standards. Until job losses began piling up in 2008, Americans were spending a greater number of hours per week at work than they had in decades, and far more total hours than do working people in other industrialized countries.

THE ASIAN OFFICE

Research bears out David Wyon's concern that improving indoor air quality through air-conditioning will be difficult in the tropics. In a Singapore call center, for example, outside ventilation improved worker output while air-conditioning did not. With ample ventilation, lowering the temperature from 76° to 72° actually reduced workers' productivity by 14 percent, but doubling the supply of outdoor air in a previously stuffy 76° office improved productivity by a whopping 35 percent.

A report from Thailand suggests that air-conditioning reinforces the need for itself. Thermal responses of workers accustomed to air-conditioned offices were compared with responses of those who normally worked in naturally ventilated spaces. Standards at the time prescribed a summer thermal comfort zone of 73° to 79°. The normally air-conditioned Thai workers found a range of 72° to 82° to be comfortable, while those who did not normally work in air-conditioning remained comfortable up to a remarkable 88°. Members of the latter group tended to wear lighter clothing, which partly explains their satisfaction with hotter conditions; however, the study's author concluded that clothing did not explain the difference entirely, that the employees also displayed "convincing evidence of acclimatization" to heat, a phenomenon to which we'll return in the next chapter.

In 2005, Japan's government set out to stop corporate wasting of energy on air-conditioning. Its voluntary "Cool Biz" campaign encouraged businesses to keep their office thermostats at 82° in summer. (The average August daily high in Tokyo is 87°.) The program can point to real successes: millions of tons of carbon dioxide kept out of the atmosphere each summer, and a widespread social stigma that is now attached to excessive cooling. Suit and tie manufacturers fought the program at first, because under it, employees were encouraged to shed their jackets and ties—the traditional uniform of the "salaryman"—

for more casual attire. They need not have worried; new markets for lighter clothing have opened up profitable opportunities. Higher thermostat settings have spread to restaurants and retail stores, but the Cool Biz program has also exposed the difficulty of maintaining both comfort and conservation in cities built for air-conditioning. Employees are now more likely to hold meetings in conference rooms where the thermostat can be adjusted downward, to bring fans to work, or even to wrap their foreheads in cooling pads. As with any temperature issue, opinions vary widely. Some employees say they're suffering, others feel good about saving energy, and a number of people say they enjoy the warmer environment. A spokeswoman for a Japanese air-conditioning manufacturer told the *Wall Street Journal* that, yes, her company too had turned up the temperature, but only to 79°. The recommended 82°, she said, "is an energy-saving environment—not a human environment. . . . Everyone became irritated and less efficient at their work." The campaign has a long way to go to achieve universal adoption, and many employees persist in wearing suits and ties.

A consumer-market analyst in Japan has argued that game theory predicts widespread wearing of business suits even at 82°. In a business negotiation, if the competitive advantage of wearing a suit when facing a casually dressed counterpart outweighs the resulting thermal discomfort, writes the analyst, employees would "rather be uncomfortable in suits than risk the penalty of showing up in 'Cool Biz' at a meeting with a suited employee from another company." A possible solution, of course, would be for participating companies to decree that all their representatives wear Cool Biz attire and for each company to publicize that fact.

COLD WAR

The U.S. military is another large sector of the economy—now surpassing more than half a trillion dollars annually—that has come to depend heavily on air-conditioning. The southern and western states form what University of Minnesota economist Ann Markusen has dubbed the "Gunbelt," a region long favored when it comes to siting of military bases and awarding of defense contracts. The value of prime defense contracts swelled by 70 to 100 percent in real terms between 1952 and 1982 in the Gunbelt, and the region has remained dominant since. Meanwhile, military contracts in the once-dominant mid-Atlantic and Great Lakes regions fell off sharply, and those on the Pacific coast stayed level. As the armaments industries moved south, they also moved toward more high-tech products, thereby becoming more dependent than ever on air-conditioning to provide a clean manufacturing environment with strictly controlled temperature and humidity.

The Sun Belt/Gunbelt went on to become the top supplier of both troops and armaments for the wars America has fought in the 1990s and 2000s. The Americans fighting and dying in the Afghanistan and Iraq wars have come disproportionately from the hometowns and military bases of the Sun Belt. Soldiers from the South may have grown up in a hot climate, but in southwest Asia, they've experienced even more extreme temperatures. In the summer of 2008, Reuters reported on the innumerable tankloads of diesel fuel that the U.S. military hauls daily by road to forward bases in Iraq and Afghanistan. Of that fuel, fully 85 percent goes to power air-conditioning systems via diesel generators. The Pentagon has struggled mightily to keep troops cool. Air-conditioning the thousands of tents in which many of the troops sleep is not exactly an efficient use of fuel, so the army has experimented with spray-on foam insulation. Vehicles, too, have been retrofitted for the heat. Canadian troops in Afghanistan found that their European-made Leopard armored vehicles became ovens in summer. The Canadian American Strategic Review noted "At best, the [Leopard's] air conditioner keeps the electronics from frying while that 'chilled' interior may be no cooler than the outside air temperature. Kandahar in the summer months hits 50°C. Tank interior temperatures . . . may reach a punishing 65°C. Urgency is indeed required; +65° Celsius [149°F] isn't uncomfortable, it's life threatening."

In both Afghanistan and Iraq, U.S. crews simply could not have survived summer after summer in enclosed, heavily armored vehicles like the Humvee were they not fitted with air-conditioning. Four years into the Iraq occupation, military historian Jon Grinspan asked whether fortified vehicles had helped or hurt troops' security. Contrasting the

Humvee with that old favorite of twentieth-century military occupa-tion forces, the Willys jeep, Grinspan remarked that in Iraq, "American troops, many military theorists now argue, are too removed in their vehicles, fighting for Iraqi hearts and minds with a drive-through men-tality. The open-air jeep meant that soldiers could, and had to, interact with the people of occupied nations; the closed, air-conditioned Hum-vee has only isolated American forces from Iraqis."

The huge frontline demand for fuel to run cooling equipment has helped create long, vulnerable supply lines in Iraq and Afghanistan, prompting the Pentagon to implement new energy-saving policies. Re-ports the *Wall Street Journal*, "If it comes to pass, the Army's planned shift away from massive, gas-guzzling supply lines carrying all kinds of luxuries to the front could herald a break with more than a half-century of U.S. military logistical doctrine."

6

SURVIVING THE GREAT INDOORS

The major objective of school room ventilation is . . . the maintenance of a room temperature of 68° to 70° with moderate air movement. . . . A minor objective should be the provision of sufficient air change to avoid unpleasant body odors. The avoidance of overheating is of primary and fundamental importance for the promotion of comfort and efficiency and the maintenance of resistance against disease.

—New York Commission on Ventilation, 1923

We don't use the air conditioner because it makes it too hot outside.

—Apartment resident, Davis, California, 1992

Seventy-four-year-old Eddie Slautas turned down his neighbors' repeated offers to install a window air conditioner in his Chicago apartment. Even when they said they'd help him pay the difference in his utility bill, he demurred. "Why should I make my electric bill higher?" he asked. "The fan is good enough." Then came a fierce midsummer heat wave. On the night of July 30, 1999, the neighbors found Slautas dead. The fan was running, blowing hot air across his body. He was one of 103 Chicagoans killed by the heat that week.

On the last night of July 2006, a Commonwealth Edison power cable running beneath the city of Chicago failed, putting 3,400 customers in the dark. The next day, as temperatures reached 100° on the fifth day of a blistering heat wave, 1,300 people had to be evacuated from high-rise residential buildings in the area. Their apartments had become saunas, so they took refuge in air-conditioned shelters. Resident Lutricia Somerville, who had resorted to spending much of the night in her parked truck with the air conditioner running, told a reporter, "It's just like Hurricane Katrina." Those trapped in the heat must indeed have felt some of the desperation that had hit New Orleans residents eleven months earlier. But the outage was short-lived, and this time no one died or suffered serious medical problems.

LIFE AND DEATH ON HEAT ISLAND

In June 2009, the U.S. Global Change Research Program—a cooperative effort by thirteen federal agencies and the White House—issued an alarming 188-page progress report on the pace of global warming. Among many dire predictions was a forecast of deteriorating human health. Thomas Karl, director of the National Climatic Data Center and the report's principal author, said health was the issue sparking the most discussion among the agencies and leading to the least certain conclusions; however, the report confidently predicted increasing rates of heat-related illness and mortality, and that higher temperatures, along with air pollution, would cause the already-accelerating rates of asthma and other respiratory ailments to rise even faster.

Heat waves continue to plague Chicago, but the city is better prepared than it once was. Its public health officials are determined to avoid a replay of the bitter experience of July 1995, when more than 550 city residents were killed by record-breaking heat. Most of those who died had no air-conditioning, or if they did, they could not afford the electricity to run it. The record numbers of air conditioners that were switched on triggered more than 1,300 failures in the electricity supply system, many of them caused by overheating of overloaded transformers. A series of blackouts hit both rich and poor neighborhoods, but the bulk of the casualties occurred in lower-income areas.

Longer, more intense heat waves hit Chicago in 1931 and 1936 but killed far fewer people. That difference between the 1930s and 1990s has puzzled experts; residential air-conditioning was virtually unheard of in the 1930s, and the inner city's population was only slightly smaller then than it is today. However, the average age of residents has increased, there is a lot more concrete to hold the heat, and analysts at the Midwestern Climate Center have suggested that people, especially older people, have become more afraid of crime and more reluctant to leave doors and windows open or to sleep outdoors (as many did in the

1930s). The analysts went on to suggest that "many people have also forgotten how to 'live and function' with high temperatures."

Air-conditioning has been credited with huge improvements in the health of the U.S. population. Ray Arsenault provided a partial list of benefits that were realized in the first few decades of climate control: "air-conditioning has reduced fetal and infant mortality, prolonged the lives of thousands of patients suffering from heart disease and respiratory disorders, increased the reliability and sophistication of microsurgery, facilitated the institutionalization of public health, and aided the production of modern drugs such as penicillin." Air-conditioning can also be an important tool in dealing with the kinds of weather crises that may become more frequent. Within the first few hours of the extensive August 2003 power blackout in the Great Lakes and Northeast, emergency rooms were overwhelmed with patients, a large proportion of them suffering in one way or another from heat stress. Most were rushed into air-conditioned shelters, where they recovered. There is also evidence that air-conditioning provides routine protection against illnesses caused by allergens, air pollution, and mosquito-borne pathogens and parasites.

Despite conflicting research results—some statistics show that air-conditioning has reduced heat-related death rates while others, as we will see, find air-conditioning's effects swamped out by socioeconomic forces—the most direct and quantifiable claim made for air-conditioning is that it can reduce the death toll during a heat wave if broad access is ensured.

The nation's average temperatures dropped following the hot 1930s, but heat waves made a comeback during the age of air-conditioning. From 1949 to 1995, the frequency of heat waves increased 20 percent, and the trend has steepened since; matters are predicted to worsen. The Global Change Research Program report, for one, is forecasting that "extreme heat waves, which are currently rare, will become much more common in the future." In a highly unusual incident that we can only hope is never repeated, an extraordinarily large, intense mass of heat and humidity in August 2003 reportedly killed 35,000 to 52,000 people in Europe. Most of the victims lived in places that normally see much milder summer weather and have few air conditioners or other means of defense against severe heat. Part of the increase in superheated weather in cities across the globe can be attributed to the heat-island effect, but these early days of global warming may already be generating more heat emergencies.

Averaged over the past century, heat and humidity have killed far more Americans than any other type of adverse weather. In forty-four of the largest U.S. cities, heat waves kill more than 1,800 people per year on average, but the annual toll rises and falls steeply, depending on whether or not there was a major-league heat wave in a given year. A 2003 study of twenty-eight U.S. cities suggests that the increase in heat-wave deaths between the 1930s and 1990s in Chicago may have been an exception; average numbers of heat-related deaths dropped by 59 percent from the 1960s and 1970s to the 1980s and continued falling through the 1990s. The study concluded, "This systematic desensitization of the metropolitan populace to high heat and humidity over time can be attributed to a suite of technologic, infrastructural, and biophysical adaptations, including increased availability of air-conditioning." A nationwide study that used statistical techniques to eliminate the effects of other socioeconomic factors found that access to central air-conditioning reduced death rates by 42 percent during heat waves that occurred between 1980 and 1985. Room air conditioners, in contrast, had no effect, except in the smallest, one- to three-room dwellings, where a window unit "may be seen as nearly equivalent to central air-conditioning."

The health benefits of air-conditioning have not been shared evenly. Historically, the most obvious disparities have been between races. In four northern cities surveyed between 1986 and 1993, 41 percent of white households had air-conditioning, compared with 16 percent of black households. Heat waves in Minneapolis and Pittsburgh killed black residents at six to seven times the rate at which they killed whites during that period (see Table 5). Almost two-thirds of the difference in heat-related deaths between the races was linked to differences in availability of central air-conditioning. Central heating didn't have the same effect; racial differences in death-rate peaks during winter cold spells were much smaller.

Table 5

Across four northern U.S. cities, the increase in mortality during hot weather was 2.5 times as large among black residents as among white residents. The comparison was between 60° and 84° apparent daily average temperature (computed as temperature adjusted for humidity). From Marie O'Neill et al., *Journal of Urban Health* 82 (2005), 191–97

Race	Increase in percentage mortality during hot weather, 1986–93				
	Chicago	**Detroit**	**Minneapolis**	**Pittsburgh**	**All four cities**
Black	5.9	12.0	17.0	12.5	9.0
White	4.1	5.5	2.3	2.0	3.7

People who enjoy easy access to central air-conditioning tend to assume that surviving killer heat is no more than a matter of motivation. One such person was Chicago's human services commissioner Daniel Alvarez. In the wake of the city's 1995 heat crisis, he told the press that its victims were "people who die because they neglect themselves." But heat kills people like Eddie Slautas not because they are lazy or stubborn but because they are under economic stresses. People with central air, who almost always survive heat waves, tend to have higher incomes; larger, newer houses; better plumbing; and higher education levels. All of those factors are also associated with lower heat-related mortality. Heat death rarely visits well-to-do neighborhoods; its victims are typically found in economically forgotten, concrete-rich, vegetation-free nooks and crannies of the larger cities. In the twenty-first century, air-conditioning has become almost universally available, yet heat waves continue to kill.

Reducing that death toll will require changing communities, not just individuals. One of several studies of the 1995 Chicago heat wave concluded that "features of neighborhoods on a relatively small geographic scale (e.g., amount of pedestrian traffic, small shops, public meeting places) affect survival rates [positively]." Marie O'Neill of the University of Michigan's School of Public Health, the lead author of one of the studies and of the research that produced the numbers in Table 5, says that while "air-conditioning is protective in the home setting," when heat waves come, home climate control is "less holistic and, of course, the more climate-damaging alternative in the long term." Both her own observations and the human experience documented in Klinenberg's history of the 1995 tragedy, *Heat Wave*, emphasize "the value of an overall healthier, more equitable, cohesive neighborhood and society for the most vulnerable residents," according to O'Neill.

Christian Warren is troubled by our dependence on artificial climate control as a remedy for ills that run much deeper: "Now you see air-conditioning pitched in the medical literature as an environmental justice issue, because it can save lives during heat waves. It has come to be regarded as another biotechnological tool. They aren't asking what really kills people. What about isolation, economic stress, crime, and paranoia about crime? You can easily imagine a couple staying shut away in their air-conditioned apartment during a hot spell, uninterested in checking on their elderly next-door neighbor, who could be dying of heat stroke."

If current greenhouse emissions continue, excess heat-related deaths in the United States could climb into the range of five thousand per year by 2050. The EPA suggests that public health officials prepare for more frequent "extreme heat events." Recommended actions include designating air-conditioned public buildings and some private buildings like movie theaters and shopping malls as cooling shelters, providing public transportation to the shelters, and (in vaguely sinister-sounding terms) targeting homeless people for "protective removal" to cooled spaces.

Cooling centers have become a common and highly effective strategy for protecting residents of big cities, not only against killer heat waves but against more routine hot weather as well. But for people already dealing with health problems, it's not easy to find the right temperature balance in a public cooling space. For example, during a mid-August 2009 hot spell, some of those taking refuge in a well-air-conditioned senior citizens' center in Brooklyn were found to be covering their shoulders with sweaters in order to stay warm. One of them, seventy-eight-year-old arthritis sufferer Vida Ebrahim, told the *New York Times*, "My apartment is so hot because the ceiling is low, so it keeps the heat. Oh my God, it's murder. But the air-conditioner, it gives me so much pain in my shoulders, in my knees."

A SHELTER FROM DISEASE, BUT A POTENTIAL SPREADER

Air-conditioning may be depended upon more heavily in the future to protect us not only against heat-induced stress but against disease as well. In the 1990s, the prospect of global warming prompted some researchers to raise the specter of more widespread mosquito-borne plagues. The World Health Organization (WHO) predicts, "Globally, temperature increases of 2–3°C [3.6 to 5.4°F] would increase the number of people who, in climatic terms, are at risk of malaria by around 3–5%, i.e. several hundred million." It's not only that the malaria season might last longer in some regions where it already occurs; mosquitoes might also carry malaria, dengue, and other illnesses well beyond their endemic areas in the tropics and subtropics when climates at higher latitudes become warmer. Focusing on the United States and other temperate areas, the federal Centers for Disease Control and Prevention (CDC) does not go as far as WHO, making no similar predictions regarding climate change and vector-borne diseases. The CDC stresses, "At this time, scientists do not have the understanding of disease ecology in each instance needed to make predictions."

The disease threat level depends as much on economic status as on geographical location. Brownsville, Texas, and Matamoros, Mexico, face each other across the Rio Grande river. In Brownsville, 83 percent of homes have air-conditioning, compared to 32 percent in Matamoros. Blood-antibody samples show that 40 percent of Brownsville residents and 78 percent of those in Matamoros have been infected by the potentially deadly mosquito-borne dengue virus. The only statistically significant differences between households that had and had not experienced dengue were associated with the presence of potential mosquito breeding environments (which were much more common in Matamoros) and access to air-conditioning. The study's authors recommended "economic assistance for air-conditioning in dengue-endemic areas."

If rates of vector-borne diseases rise, most of the new illnesses will come in the world's less privileged countries. In America, roomy air-conditioned houses already offer protection against the bites of garden-variety mosquitoes; if there's an onslaught of their exotic, pathogen-laden cousins, those same protections will be effective. Since the 1980s, residents of the southeastern United States have found themselves driven increasingly indoors by the Asian tiger mosquito. The small striped import is a terrific annoyance, administering its unusually irritating bites not only in early mornings and evenings but throughout the day. It can be a carrier of dengue, eastern equine encephalomyelitis, and Cache Valley viruses. The species is expected to spread northward with any overall rise in temperatures.

Paul Reitner, who studies vector-borne diseases at a CDC laboratory in Puerto Rico, has thoroughly reviewed humanity's past experience with malaria, dengue, and yellow fever. Relying on that history to weigh our prospects in a warmer future, he concluded in 2001 that the "changes in lifestyle and living conditions" that have prevented large malaria outbreaks in countries with temperate climates will continue to give adequate protection as those climates warm. Vaccination, he predicted, will take care of yellow fever. As for dengue, the United States can be expected to remain safe, wrote Reiter, but the warmer Mediterranean countries could find themselves facing dengue outbreaks if the mosquitoes that carry it return in force. Air-conditioning use has grown fast in the Mediterranean region since his paper was published, but it is still far less common there than in the southern United States, and windows are kept open for most of the summer in the densely packed cities of southern Europe and North Africa.

In California, the potent combination of two technologies, air-conditioning and television, has already helped keep in check two other mosquito-borne diseases: western equine encephalomyelitis and St. Louis encephalitis. Surveys in the 1980s showed that on hot summer evenings, people in the state's central valley tended to stay indoors and watch TV during the very hours when the vector of the two viruses, a species called *Culex tarsalis*, tends to feed. More recently, in 2006, the CDC also recommended air-conditioning as a preventive measure during the summer months when several *Culex* species that carry the West Nile virus are most active.

While air-conditioning helps isolate humans from some pathogens, it can encourage infection by others. The moisture that cooling systems circulate or wring from the air can provide ecological niches in which diverse gardens of microorganisms thrive. According to a 2003 paper in the *Lancet*, "Heavy growth of bacteria, fungi, and protozoa has been documented in air-cooling units, air-conditioning cooling coils, and drip pans within office buildings. Microbial contamination has resulted in outbreaks of rhinitis, humidifier fever, asthma, hypersensitivity pneumonitis, and Pontiac fever." The study also reported on the effects of irradiating the coils and drip pans in three office buildings' cooling systems with ultraviolet (UV) light. Irradiation dramatically shrank populations of bacteria and fungi and reduced the occurrence of illness symptoms among workers. Most of the health improvement may have come from a reduction in allergic-type reactions to the microbes rather than from lower infection rates. The authors estimated that installation of UV equipment in most North American office buildings "could resolve work-related symptoms in about 4 million employees, caused by microbial contamination of heating, ventilation, and air-conditioning systems."

The bacterium *Legionella pneumophila* made headlines in 1976 when it killed thirty-four people who were attending an American Legion convention in a Philadelphia hotel. After a six-month investigation, health officials announced that the pathogen had invaded the hotel via the air-conditioning system's cooling tower, which offered a nice, moist growth environment. The previously unknown bacterial species was named in memory of the more than two hundred convention-goers and hosts whom it had sickened or killed, and the form of pneumonia it causes was designated "Legionnaires' disease." The largest outbreak to date occurred in Murcia, Spain, in 2001, when an estimated 450 to 700 people were infected by *Legionella pneumophila*. The bacteria were found to have come from a *hospital*'s cooling tower. The disease continues to strike thousands of Americans each year and can have a fatality rate of 5 to 30 percent in serious outbreaks.

A 2004 National Institutes of Health review concluded that the disease "might not have occurred were it not for the environmental niche provided by air-conditioning systems."

MOLECULES AND THEIR SUMMER MEMORIES

Unfriendly microorganisms may not be the only health hazard of the air-conditioned life. Basic research in biology is showing that by dramatically altering our indoor environment to fend off climatological and biological threats, we may be changing the ways in which our own bodies function. As we primates evolved in hot climates, nature equipped us with heat-adaptation mechanisms. Suffering through high temperatures, especially when we are also exerting ourselves, builds up tolerance to heat. After days or weeks in the heat, our sweat glands swing into action more promptly and increase their output when heat hits; in addition, blood volume increases, blood flow is redirected to skin vessels to dissipate heat, the heart rate stays lower and steadier, the body's core temperature lowers, and we are able to function and work for longer periods under hot conditions. Previous exposure to heat also allows the body's core temperature to rise without causing stress. Marathon runners, for example, can tolerate higher body temperatures than their fellow mortals.

It has been known for a couple of decades that "heat-shock proteins"—molecules that the cells of virtually all organisms can produce rapidly in response to high temperature—help protect the body's individual cells from being killed by extreme heat. In recent years it has also become clear that with exposure to heat, these proteins accumulate in the body's cells and help it to achieve quicker adaptation to subsequent heat stress. If the weather cools or you manage to stay out of the heat for an extended period, the acclimatization gradually wears off. But it has been reported that when laboratory rats are acclimatized through heat exposure, then allowed to enjoy two months of cool "de-acclimatization" before being reexposed to heat, their bodies reacclimatize much more quickly than do those of rats who never got the first heat treatment. This rapid response of heat-adapted rats to subsequent heat—a response that has also been seen in humans—has been attributed to "molecular memory" involving production of heat-shock proteins. In the original heat-stress hypothesis, heat shock proteins served as molecular versions of emergency medical technicians (EMTs), showing up within minutes to save the lives of heat-crippled cells. With further research, it has become clear that they are indeed deployed by the body as EMTs, but that they may hold a second job, that of personal trainers who help the whole body respond better the next time it is challenged by heat.

Research, much of it done by the U.S. military, has shown that humans reached about 75 percent of maximum adaptation to heat after about five days of controlled exposure; after ten days, the process is almost complete, but benefits can continue to accumulate for up to three weeks. One hour of exposure per day does not appear to do the trick, but two hours seems to be more than enough. The daily exercise routine through which researchers typically put their subjects includes simple activities like walking on a treadmill in a room where the temperature is in the high nineties or low hundreds (with or without high humidity). Such techniques are regularly used as part of military or occupational training to prepare people quickly for working in the heat. Less strenuous exertion in the heat can also lead to acclimatization, but it takes longer and may not reach the maximum level. And the exposure should be near-daily; intermittent hot periods are far less effective. Dr. Paulette Yamada, a heat-stress researcher at the University of California at Los Angeles, says that repeated exercise in the heat helps a person achieve the physiological changes needed for heat adaptation, and "I don't think a weekend warrior who exercises twice a week would reach the same level of heat adaptation as a person who exercises five to six times per week."

Based on this research, we can expect people who live with the heat or whose jobs or pastimes involve regular heat exposure to be better prepared to deal with hot weather than are those who spend most of their time in controlled climates. Michal Horowitz, professor at the Hebrew University of Jerusalem, and her colleagues have done much of the pioneering research on the biology of heat stress and acclimatization. She confirms that "those who live exclusively in an air-conditioned environment endanger their ability to cope with severe heat load."

SICK BUILDINGS

The earliest struggles over air-conditioning and health were focused not on private homes but on workplaces and schools. There was no doubt that when people congregated in enclosed spaces for long periods, they tended to suffer health problems, and that fresh air could have an invigorating effect. But what was wrong with the indoor air? Eighteenth-century theories pointed to a buildup of carbon dioxide and later to unidentified "crowd poisons." Whatever it was that accumulated, whether it was particulate, chemical, or biological, the solution seemed obvious: dilute it. Engineers, architects, and builders of the late 1800s and early 1900s became expert in mechanical ventilation of large spaces, and governmental bodies set standards for the number of cubic feet of outdoor air that had to be flushed through a space each hour. The ultimate goal was to keep people healthy, but the immediate target was the overwhelming blend of bodily odors that tended to accumulate in any closed, crowded space, especially in summer. Gail Cooper relates how one New York cinema of the pre-air-conditioning era was "so vile that an attendant walked up and down the aisles with an atomizer of perfume vainly trying to mask the odor of the theater crowd."

By 1911, the scientific community was changing its collective mind, deciding that people were not suffering because of something they inhaled from the indoor air but simply because the temperature and humidity of the air reach unhealthy levels in enclosed, populated spaces, and that being uncomfortable makes people feel sick. The American Society of Heating and Ventilation Engineers (ASHVE, predecessor of ASHRAE) and its members' companies responded in a flash. Engineers went to work, intent on identifying and creating the ideal indoor climate, one that never becomes too warm or too cool, never too dry or too humid. That quest eventually led to modern central air-conditioning systems.

Arguing for the opposite approach, writes Cooper, were the open-air crusaders: "Composed of school officials, public health professionals, and social reformers, this alliance advocated the maximum exposure of school-age children to the healthful influence of the outdoors." Some took the crusade beyond the limits of good sense, leaving classroom windows open all winter, with just enough heating to keep the temperature above 40°. The open-air crusaders won a fleeting victory when, in a thick 1923 report, the high-profile New York State Commission on Ventilation came down on the side of open windows. But by then, ASHVE had beaten the commission to the punch with its publication of the first psychrometric chart, called the "Comfort Chart," which illustrated the ranges of indoor temperature and humidity that would keep people of all ages comfortable and, presumably, healthy. Precise and scientifically framed, the Comfort Chart eventually won out over open windows, and thus climate control beat out fresh air. Decades would pass before air-conditioning managed to penetrate every corner of American life, from the classroom to the bedroom to the football stadium, but with the Comfort Chart, industry had laid the philosophical foundation for its growth.

Once air-conditioning became the norm in large buildings, doubts began to surface. Through the 1980s and increasingly in the 1990s, many office workers and their doctors began to argue that the pendulum had swung too far, that buildings tightly designed for maximum year-round comfort and energy efficiency were actually making people ill. Eighty years after the New York State Commission drew attention to the need for ventilation in schools, air quality in the classroom was once again a hot issue. Compiling what was known in 2003 about the effect of indoor air on the health of schoolchildren, a team of U.S. government and university scientists concluded that "ventilation is inadequate in many classrooms," many of which by that time were air-conditioned. Asthma and other respiratory problems were widespread, and volatile organic compounds, molds, and allergens in floor dust were prime suspects. The spread of air-conditioning has been associated with the spread of what we now know as sick building syndrome. Studies in California, Brazil, France, the United Kingdom, Denmark, and elsewhere showed that people employed in air-conditioned workplaces had poorer health and spent more time in physicians' offices and hospitals. In a commentary that appeared alongside the French report in a 2004 issue of the *International Journal of Epidemiology*, independent expert Mark Mendell wrote,

Occupants of office buildings with air-conditioning systems (e.g. central ventilation with cooling of air) consistently report, on average, more symptoms in their buildings than do occupants of buildings with natural ventilation. This has been the finding in individual studies from many studies over the last 20 years, and in three reviews. The symptoms in these studies have included mucous membrane irritation, breathing difficulties, irritated skin, and constitutional/neurological symptoms such as headache and fatigue.

The French study monitored the health of 920 women who were enrolled in an experiment to test the efficacy of antioxidant nutritional supplements. They were administered a questionnaire that included the question, "Is air-conditioning in use in your workplace?" Those who answered "yes" were 70 percent more likely to have work absences due to illness and were 50 percent more likely to have been hospitalized. Mendell acknowledged that several factors might have skewed the results. For example, the women in higher-income jobs would be more likely to have both workplace air-conditioning and better medical care, and warmer cities where more offices are air-conditioned may be more disease-prone. Mendell argued that any such sources of error would be minor, however, because nationalized health care in France tends to level out class differences in access to medical treatment and the study attempted to adjust statistically for regional differences.

In buildings with low ventilation rates, incidence of respiratory illness is increased by 50 to 120 percent, according to one review of the health literature. Most of what we know on the subject comes from commercial and public buildings; the effects on families when residences are tightly sealed year-round are much less well understood. In a 2003 review, medical researchers at Johns Hopkins and Harvard universities commented that "remarkably, there has never been a comprehensive study on the role of ventilation and health and comfort in homes." Home air-conditioning systems, which generate condensation but generally do not involve circulation of cooling water, are less likely to have bacterial cultures multiplying within them than do systems in larger buildings; however, inadequate ventilation could lead to the same kinds of respiratory problems that have been associated with poor ventilation in schools. Although ASHRAE recommends that, each hour, 35 percent of the air inside a home be replaced by outside air, the researchers reported that "associations of home builders have resisted attempts to specify mechanical means to achieve this recommended exchange rate or to institute higher exchange rates for homes."

AN APPETITE FOR COMFORT

Air-conditioning may also play at least a small role in another trend affecting our health: obesity. The number of Americans who are overweight has shot up dramatically since around 1970. Just in the 1990s, the rate of abdominal obesity rose by 67 percent in children and teenagers. Most analysts have attributed climbing obesity rates to the aggressive marketing of calorie-rich food and drink, coupled with a decline in physical activity. But in 2006, a team of twenty medical researchers across the United States published a paper exploring ten other "equally compelling" explanations for the nation's obesity "epidemic." Along with sleep loss, endocrine disruption, decreased smoking, aging of the population, increases in age at childbearing, effects of the fetal environment, pharmaceutical use, natural selection, and mating preferences, they listed "reduction in variability of ambient temperature" as a potential villain. Because of central air-conditioning and heating, they wrote, today's population spends much more of the year in environments that are within the thermoneutral zone (TNZ)—the range of air temperatures at which the body doesn't have to expend energy to maintain its normal internal temperature of 98.6. When the surrounding temperature rises above the TNZ (when it goes above the seventies if you're wearing normal clothes, above the eighties if you're naked), the body has to work to rid itself of the excess heat generated by normal metabolic processes; below the TNZ, increased cellular metabolism and more muscle action, including shivering, tend to burn energy. When we're feeling too hot or too cold, we use up more of the chemical energy provided by food and convert less of it to fat. The authors concluded that while no one would seriously suggest that more Americans take up smoking to lose weight, "factors such as sleep reduction and increased use of heating and air-conditioning might be things that are easily modifiable and for which modifications in the direction that would hypothetically reduce obesity levels would also have added benefits (e.g., a more healthy and alert population and less use of fossil fuels)."

When the weights of actual twenty-first-century human bodies are analyzed, it is impossible to separate the interrelated impacts of air-conditioning and overeating. Decades of evidence suggest that we tend to eat more in a cooled environment than we do when we're hot and sweaty. Military officials were the first to examine this issue because they were concerned that overheated troops might fail to eat enough to support heavy exertion. Putting it as directly as possible, one researcher wrote in the 1970s, "Cattle, swine, rats, goats, and U.S. Army men all eat more when the temperature is low than when it is high." Indeed, livestock trials have shown that keeping either the housing or the feeding environment at a comfortable temperature increases food intake. Animals that lived under warm conditions but were led into a cooled room to feed had a bigger appetite than ones that dined in the heat. In a study of indoor air conditions in a Danish school, children reported being "more hungry" when classrooms were kept cooler (they also performed better in three out of eight academic tests when temperatures were lower). Another military study cited a survey of restaurants in the Toronto area showing that not only did business decline in the summer, but the average purchase per customer also decreased by 2 to 20 percent. Restaurant owners reported that sales plunged much deeper during heat waves, but after spending a while in the air-conditioning, customers tended to order normally. Sales dropped dramatically when cooling systems broke down.

A 2006 paper published in the journal *Economics and Human Biology* reported that for most of Tokyo's elementary school students, body mass varied normally with the calendar, with weight gains in winter and losses in summer. However, obese children displayed a wholly different pattern, gaining weight in summer. The authors, Drs. Masako and Maiko Kobayashi, interpreted the striking results:

> Before the 1970s, it was quite natural for Japanese to lose weight during the summer. However, the air conditioner changed their

lifestyle beginning about 30 years ago. A total of 5.7% of Japanese households had an air conditioner in 1970. Since then, the percentage has increased annually. In 2004, the percentage reached 87.1% overall and 99% in Tokyo. . . . If children could stay comfortably indoors most of the summer holiday, playing computer games, watching TV or studying without exercise, it seems that they are able to maintain a good appetite and gain weight.

Perhaps the most significant, if hard to quantify, link between air-conditioning and obesity is through lack of exercise. The decline in the outdoor life described in chapter 3 has meant a decline in physical activity as well. Even without air-conditioning, children and adults would almost certainly watch TV and sit at the computer in summer, but probably not for hours at a time.

The dominance of indoor life may also be contributing to vitamin D deficiencies. Once thought to be a problem of the past thanks to supplementation of food, deficiency of vitamin D has returned with unexpected force. A 2009 report showed a sharp drop in Americans' average blood levels of the vitamin between two surveys done in 1988–1994 and 2001–04. Three out of four adolescent or adult Americans were found to suffer from insufficiency. The human body makes the bulk of its own vitamin D, but requires exposure of the skin to sunlight in order to do so. Researchers worried that extensive use of sunscreens was blocking ultraviolet-B rays, which are essential to the process. Previous research had also shown a strong link between a person's extent of outdoor activity and blood level of the vitamin.

On another front, with 64 percent of Americans complaining that they suffer from chronic sleep problems (41 percent report that they have problems almost every night), any product or practice that improves the chances of a good night's sleep could improve the country's general health picture. By moderating heat and humidity and helping shut out noise, dust, pollen, and other irritants, air-conditioning has become one of the country's top nonpharmaceutical sleep aids.

But sleeping in cool comfort may take a physical toll. In summer, people often complain that going back and forth between chilly air-conditioned buildings and the outdoor heat makes them "feel bad," even ill. A study in Japan, published in 2006, identified a possible chemical contributor to that phenomenon. Cortisol is a hormone produced by the adrenal glands that is a key part of an endocrine apparatus regulating the body's metabolism, blood flow, immune responses, nerve firing, and other functions. Normally, wrote the study's authors, the hormonal system in which cortisol plays a central role "is of major importance with regard to an organism's response to physical or psychosocial stimulation, and morning awakening should be associated with bursts of cortisol secretion." In an experiment conducted in September, that morning cortisol "burst," which normally hits our bloodstream twenty to forty minutes after we wake up, was delayed by an average of two hours in people living in an air-conditioned home all summer. The previous June, before those subjects had spent much time in air-conditioning, they experienced the same normal cortisol burst as did people in non-air-conditioned homes. The authors concluded that although air-conditioning can "reduce physiological stress from the summer heat," the big fluctuations in temperature and relative humidity that the human body experiences in going in and out of air-conditioning during the day could actually cause stress and disrupt cortisol rhythms. The study's results suggest that weeks of exposure to air-conditioning can delay the daily release of cortisol until 10:00 A.M. or later, and that just might turn a morning person into a slow-starting sleepyhead.

When faced with the summer onslaught of immune-system irritants, millions of allergy and asthma sufferers take refuge indoors. Yet the air-conditioning era has seen allergies become an ever-larger problem in Western societies, and the prevalence of asthma has doubled with each decade that passes. The still-evolving "hygiene hypothesis," first articulated in 1989 by British epidemiologist David Strachan, says that the immune systems of allergy and asthma victims have been disoriented in part by insufficient childhood exposure to bacteria, fungi, nematodes, and/or other tiny organisms. An earlier form of the hypothesis blaming a reduction in childhood illnesses is now outdated; rather, it appears that youngsters are not getting enough exposure to the more benign microscopic inhabitants of the outdoor environment, especially the rural environment.

As kids lounge in cool, sterile interiors—whereas in the past they might have been in the backyard making (and maybe tasting) mud pies—they may be predisposing themselves to allergies and asthma. The mechanisms are exceedingly complex and still not fully understood. Recent work suggests that when a person's body is exposed to a wide array of harmless microorganisms early in life, its immune system "learns" to restrain itself, probably through the action of bodies called regulatory T-cells. Without such exposure, the system habitually overreacts to noninfective organisms and particles, going on the attack when no attack is called for. To the extent that air-conditioning limits children's exposure to the microbe-rich summertime landscape by keeping them behind closed doors and windows, playing in the filtered air, it could be depriving their immune systems of essential training. However, asthma is also increasing in inner cities and in urban areas of impoverished countries, so protection probably isn't induced by exposure to just any old microbes. Simply living in a dirty or clean city doesn't guarantee that a child will or won't have problems, and the seemingly strong protection provided by farming environments still has not been fully explained. Until much more research is done, the deep details of the hygiene hypothesis won't be fully understood, but in the meantime, kids out of school for the summer would be well advised to go out and play.

Meanwhile, evidence is emerging that nematodes—microscopic worms of many species, some of which cause serious human diseases—produce molecules that keep the immune system calm, and the absence of exposure to them may play a role in the widespread occurrence of inflammatory diseases in industrialized nations. Graham Rook of the University College London Medical School lists several disorders that are increasing in frequency, and that, based on epidemiological associations and evolutionary logic, are likely to be exacerbated or made more common by lack of exposure to nematodes: cancer, atherosclerosis, Alzheimer's disease, and even clinical depression and anxiety. Rook concludes that although much of recent human evolution has been shielded from natural selection because we "easily detect problems with the physical environment and invent appropriate technological adaptations," our immune systems may have continued evolving during the early millennia of agriculture and settled life. Now, with our species' almost instantaneous shift out of nature into a human-built environment, immune systems may not be receiving the appropriate inputs. Specifically, our bodies may have declared a truce with some invaders, including so-called "old friends" like *Lactobacillus* bacteriae and old enemies like pathogenic nematodes (the latter because "although not always harmless, once they were established in the host any effort by the immune system to eliminate them was likely to cause tissue damage"). The truce meant restraining the immune system in the presence of such organisms, with the result that we became dependent on those same organisms to keep our immune systems from overreaching. In a protected environment, immune systems can go haywire. Although we should heed the warning that "it would be a dangerous and Western-centric jump from [the nematode-immune system link] to nostalgia for parasitic infestation" like those that plague millions in poor nations, emerging evidence raises some weighty questions about increasingly common immune system disorders and the increasingly sterile environment in which we spend most of our time.

THE BEDROOM CLIMATE

Wherever room air conditioners have seen increasing sales, starting in 1950s America, the first place they have been installed is the bedroom, and their buyers often have been shopping for more than a good night's sleep. Although data are surprisingly sparse, it is generally accepted that sexual desire is enhanced when the parties involved are not uncomfortably hot.

With that in mind, a prominent air-conditioning manufacturing company in Hyderabad, India, ran a quarter-page newspaper advertisement in the late 1990s to plug its new generation of high-performance room units. The ad featured the faces of a man and woman in a reclining position and soft focus with a caption that read "... for improved performance in the bedroom." In sexually conservative south India, the ad drew a storm of complaints from outraged readers, but the double entendre gave sales the big boost that company executives knew it would.

Although it's an imperfect indicator, one way to track the link between climate and sexual activity is to examine seasonal variation in births. Human birth rates fluctuate predictably with the calendar, with different patterns in different regions of the globe. Before 1930, when outdoor work was much more common that it is today, annual variation in hours of sunlight had the strongest influence on the seasonality of births, with temperature a secondary influence. As the effect of day length was washed out by indoor work and artificial lighting, temperature became the dominant influence. Now the seasonal impact of temperature has been partially blunted by air-conditioning.

American maternity wards see their largest numbers of vacant beds in May, nine months after the midsummer temperature peak. A statistical summer "trough" in conceptions is attributed primarily to couples' slackened interest in sex, with a secondary possibility that greater heating of testicles leads to lower sperm viability. This seasonal pattern is strongest in the southern United States and India but does not occur at all in northern Europe, where summer temperatures are not as high. Demographic studies shown that in most parts of the United States, a 10° increase in average monthly temperature results in a 4 to 10 percent reduction in conceptions.

The seasonal fluctuation of sexual activity and conception has weakened over the past century, however, and air-conditioning has played a key role. The effect of air-conditioning in dampening the seasonality of births in states across the United States has proven much stronger than the impact on month-to-month birth rates of other socioeconomic factors such as income and education. Some of the most convincing evidence comes from racial differences: seasonality didn't decline until later among America's nonwhite people, and the lag closely parallels the lag in nonwhite families' gaining access to air-conditioning.

I have seen no evidence that air-conditioning leads to an increase in *overall* birth rates. Simply flattening out the seasonal variation in sexual activity and births without affecting population growth has no obvious environmental impact. But it does provide one more illustration of the power that a human invention like air-conditioning has to mold the biological patterns of human life.

7

INDIA: WHERE "A/C" MEANS "VIP"

*Think about it: fifty-six million people displaced by Big Dams in the last
fifty years. . . . When the history of India's miraculous leap to the forefront of the Information Revolution is written, let it be said that fifty-six
million Indians (and their children and their children's children) paid
for it with everything they ever had. Their homes, their lands, their languages, their histories.*

— Arundhati Roy, *Power Politics*, 2001

In India, more people spend more hours each year in extreme heat
than in any other nation. Although foreign residents and visitors discuss the heat obsessively, people who grew up with India's seasonal
cycles and without artificial cooling tend not to talk much about their
thermal plight, at least until temperatures rise so high that thermom-
eters start to fail. Extreme heat descends somewhere on the subcontinent every summer; an extreme among extremes occurred in the
southern state of Andhra Pradesh and its capital Hyderabad in May
2002, when daily highs in some localities exceeded 120° several days in
a row, hundreds of people died, and, according to news reports, "birds
fell dead from the trees." It wasn't an entirely natural disaster. Heat in
the state is aggravated by deforestation and falling levels of the state's
reservoirs and other surface waters. Fierce heat waves hit various regions of the country regularly, exacerbated not only by shortages of
trees and water but also by the increasing volumes of concrete and rock
concentrated in urban areas.

It may be worsening, but spectacular heat is an old story in South
Asia. In the mid-nineteenth century, Reverend J.M. Merk, a sixteen-
year resident of British India, provided this vivid description of
Punjab's summer environment:

A denizen of the temperate zone can hardly realize to himself the desiccating, truly scorching heat of this wind. When exposed to it, one may imagine one is facing an open furnace. The thermometer rises in the shade to over 50°C (122°F). . . . At sunrise, or soon after 5 A.M., houses must be closed, only a small door being left open for communication with the outside. Thus the house of a European is more like a gloomy prison than an ordinary dwelling-house. So long as the hot winds blow strongly and steadily, rooms may still be kept in some measure cool by means of *tatties* or grass screens set in front of the doorway, and occasionally sprinkled with water, or by the fan vanes of the so-called "thermantidote," which a servant keeps revolving and sprinkles with water; and at night the *punkah* [a fan, also servant-powered] is worked. Whoever cannot provide himself these artificial cooling appliances must suffer the daily torment of insupportable exhausting heat. Man and beast languish and gasp for air, while, even in the house, the thermometer stands day and night between 35° and 45°C (95° and 113°F). Little by little the European loses appetite and sleep; all power and energy forsake him.

But when the winds finally calm, warned Merk, "now indeed the heat is truly fearful." And when the monsoon rains finally descend, bringing more moderate temperatures, "the atmosphere weighs on one like a heavy coverlet. . . . Shoes and all articles of leather become thickly coated with fungus, books become mouldy and worm-eaten, paper perishes." When the clouds lift in September, "the heat soon becomes once more so great that one longs for the cold season." Finally, tolerable weather arrives sometime in November, but by March, "the hot summer is at hand" once again.

The monsoon season brings deliverance from the worst of the heat, but, more important for India's thirsty majority, it replenishes water supplies that receive no input of rain for seven to nine months of the year. In summer, the water equation is squeezed at both ends: hydroelectric plants can't sustain their share of the electrical supply, and more and bigger appliances in the cities force demand to new heights every year. But those who can afford air-conditioning can also ensure access to water. For many of them, heat has come to rank as the bigger issue. As a marketer for LG Electronics told Reuters in 2002, "The best attribute of an A/C is its addiction." A spokesperson for competitor Samsung added, "[It] has moved from its luxury status to a necessity item just like any washing machine or refrigerator."

POWER TO WHICH PEOPLE?

By 2020, energy consumption by air-conditioning in India is projected to grow almost tenfold compared to its 2005 level. That would overwhelm any efficiency improvements (see Figure 3) and push energy de-

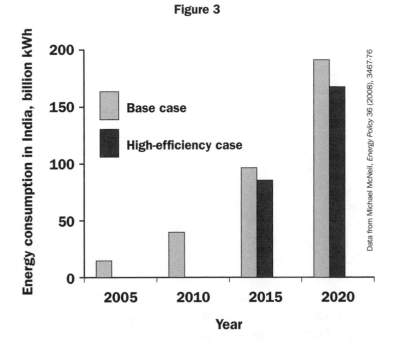

Figure 3

Total estimated electricity consumed annually in India by air-conditioning systems under a "base case" (in which equipment efficiency remains constant) exceeds that under a "high efficiency case" (in which efficiency of the average unit improves by 16 percent), but not by much. The estimate for 2005 is based on known numbers and types of units, and estimates for other years are projections by Lawrence Berkeley National Laboratory researchers.

mand for cooling to a level that puts India in the same league as Western nations. Although many more households own refrigerators than own air conditioners, total energy usage for air cooling was already double that for food cooling in 2005, and could rise to quadruple the level of refrigerator consumption by 2020. Despite the fact that central air-conditioning is rare in Indian homes and room units are run sparingly, they use much more energy than the far more numerous refrigerators.

India's power generation capacity is being expanded year by year but always lags behind demand; 2008 was typical, with shortfall in total supply approximately 10 percent. Eighty-two percent of the urban population has access to electricity, but outages and brownouts are routine. Wealthier urban areas tend to see fewer and shorter power cuts than do downscale neighborhoods or slums. Out in the villages, where a large majority of Indians live, only 42 percent of homes have any electrical service. In those rural areas, the only predictable features of the electricity supply are voltage fluctuations and daily blackouts; people tend to speak of the number of hours "the current" is on, rather than discuss the much greater length of outages. Cuts occur at all hours, often unpredictably. Villagers or urban dwellers lucky enough to have a ceiling fan can find themselves spending the better part of a summer night sweating, staring through the dark at the motionless blades, silently begging the fan to come back to life.

India generates about 55 percent of its electricity from coal and another 25 percent from other fossil fuels. Planned efforts to use less coal and more natural gas will make electrical generation more costly, and that will put utilities in direct competition with manufacturers of precious nitrogen fertilizer, a product that currently accounts for 40 percent of the country's natural gas consumption. Electricity generation in India is expected to go from its 2003 level of 567 billion kilowatt-hours (473 total kilowatt-hours per capita, compared with more than 1,650 kilowatt-hours per capita just for air-conditioning in the United States) to 1.2 trillion kilowatt-hours in 2025.

Much will depend on how that doubling is achieved. New fossil-fuel and nuclear plants draw opposition similar to that which they face in the United States. Twelve percent of India's electricity comes from hydroelectric plants, and rapid growth of that source is being looked to

by officials as a way to meet the country's rapidly expanding power demand without burning more fossil fuel (except, of course, during dam
and power plant construction). Dams have a controversial history on
the subcontinent. Estimates of the number of Indians flooded off their
lands by big hydroelectric projects run into the tens of millions. The
fiercest and most protracted battle has been over a series of dams on the
Narmada River in the country's west that have displaced an estimated
350,000 people from their land and wiped out a vast refuge for biodiversity. The largest and most contentious of the dams is the one nearest the river's mouth—the Sardar Sarovar project. A report issued by the
Prayas Energy Group states, "If we want 40 percent of electrical generation to come from hydropower (as reported in the official literature),
then we need to build new dams producing nearly 4,000 megawatts
each year. This is equivalent to *two* Sardar Sarovar projects each year!"
Further construction of dams for water management, power generation, or both will be extremely difficult politically, judging from recent
history. Popular protests against further flooding of villages, farms, and
natural lands have been massive, vigorous, and unrelenting.

The 3 percent of India's electricity that comes from nuclear plants
will likely grow, but not enough to become a major source. Girish Sant,
who analyzes India's persistent power shortages for Prayas, suggests a
different approach: "One needs to ask, 'Why allow air-conditioning at
all in such an energy-short situation?' We should tell people, 'Look, if
you want to have air-conditioning, buy your own photovoltaic solar
array.' The rich can certainly afford to do that."

Unless and until that happens, most of the country's electricity will
continue to be generated with fossil fuels. As a substitute for onerous
and degrading manual labor, fossil energy does have its progressive
side, however—a side that's on display in India as much as anywhere.
With electric fans used by 77 percent of urban households and universal in upscale houses, hundreds of millions of people can stay as comfortable in summer as did a maharaja or British colonel of yesteryear
without ordering a servant to turn a thermantidote or pull on a punkah
rope all night.

To take the idea further, consider a household lucky enough to be
among the fewer than one in fifty that have air-conditioning. Its occu-

pants experience a level of cooling that, if human powered, would require an outrageously large workforce. A human pedaling a stationary bicycle steadily at the equivalent of twelve miles per hour can generate about seventy-five watts. To run a typical 1,800-watt Indian room air conditioner would then require the average work output of twenty-four people. (For highly efficient equipment, the figure would go down to "only" twenty people. Still, those people would be condemned to working in the heat. If they worked in an air-conditioned room, the power they generated would be just about enough to cool themselves and the room, with none left over to keep their employer cool.) Following the logic that says economic growth through labor-saving technology makes for a more humane society, the air conditioner is far more humane than the electric fan, because, theoretically, it relieves not one but twenty-four people of demeaning physical labor.

COMFORT VERSUS FOOD

From air-conditioning's tentative introduction in the late fifties until the early 2000s, its growth in Indian society moved at a glacial pace. Evaporative cooling remained dominant in dry summer environments across India. Until recently, Darshan Bhatia co-owned and co-ran a family air-conditioning company founded in Hyderabad in the 1950s—the same company that ran those provocative "bedroom" ads in the nineties. He told me about growing up in the 1960s and 1970s: "My father owned an A/C company, yet even I grew up without it. There was an idea, which persists today, that to give children air-conditioning in their bedroom would be to spoil them. The only unit we had at home was in my parents' bedroom. In fact, early on, my dad had a very small unit that he inserted into a hole in the mosquito netting that hung over my parents' bed. So it was kind of like an air-conditioned tent that held the cool air closer around the bed a bit longer."

For many years, even in wealthy families, Bhatia told me, air conditioners have been installed almost exclusively in the master bedroom. Typically, even that one air conditioner has to wait its turn behind other big purchases. As incomes rise, says Bhatia, families invest first in cell phones, then in a refrigerator, then in a television set (because there is great demand from the whole family for all three of those items), then a washing machine (to reduce outlay for domestic labor), and finally in an air conditioner. As air-conditioning use has expanded in recent years, he says, "The second unit will usually go into the room of a grandparent or other elderly person. Third would come kids' bedrooms, and fourth, the living room." But, he says, very few living rooms are air-conditioned even today, and an air-conditioned kitchen is virtually unheard of. "In a small business today, the boss typically will have an air conditioner in his office, but that would be it. However, many of the newly built, large office buildings are being centrally air-conditioned."

For a modestly well-off family, "buying an A/C unit is generally not a problem," Bhatia says. "It's the running cost that's high." With a ton-and-a-half unit selling for around 25,000 rupees (a little over $500 at the current exchange rate) and with Bhatia's estimate of ten rupees per hour for the electricity to run it, a Hyderabad home owner who wanted to run a single unit for 2,500 hours annually (as do typical homeowners in Florida), the operating cost would surpass the original cost of the equipment within the first year. So most units are run not continuously but only when the room in question is occupied. But, says Bhatia, "It's kind of like a cell phone. When you first get it, you might say, 'I'm not really going to use this much, just for emergencies or for a couple of hours at night in the summer.' But before you know it, you depend on it." It feels so good, and is so effective at blocking out the increasingly high levels of dust and noise pollution that plague India's cities today, that people keep units running much more than they initially intended, if they can afford to do so.

Most Indians are more sharply aware of the fundamental fragility of food production than are most Americans, and a large majority still live in rural villages. By necessity and by law, the water pumps that serve the nation's rice paddies and other irrigated fields take their share of the electrical supply right off the top. As of the year 2000, the farmers of Andhra Pradesh were consuming about 40 percent of the electricity produced in the entire state. The Punjab state government has made its priorities clear, imposing a temporary ban on the use of air-conditioning in 2004 in order to give their pumps enough power to save the state's rice crop. Four years later, power-short farmers in Ma-harastra organized a protest against chronic shortages at the private residence of the state's chief minister. They became far more vocal when they noticed multiple air conditioners attached to the house.

In the southern state of Tamil Nadu, domestic electricity use, which had accounted for less than 10 percent of total electricity consumption in 1980, now exceeds 25 percent. Annual sales of air conditioners shot up 300 percent between the early 2000s and 2008. In 2008, the state's electric power regulators were seeing a sharp increase in residential customers' bills, above a threshold that "would generally indicate that air conditioners are being used." Although high-consuming customers pay higher rates per kilowatt-hour—a practice common in India—even those higher rates don't bring in enough revenue to cover what it costs the government to produce the electricity; therefore, even the high-end customers are subsidized. Farmers and industrial electricity users have joined in protest, according to the *Hindu* newspaper, "voic-ing demands for enforcing discipline on the domestic consumers, at least high-end consumers." State electricity board officials instead initi-ated a program to encourage consumers to buy more efficient air con-ditioner models.

Energy efficiency, however, can be a slippery issue. Darshan Bhatia warns that "there is probably nothing as easy to cheat on as is efficiency. Suppose you are advertising a one-and-a-half ton unit that uses 1,800 watts. You're saying the energy efficiency rating (EER) is 10 [18,000 Btu/1,800 watts], which is good." But in practice, he says, that claim is often an intentional lie; the unit may actually put out only 15,000 Btu, making the real EER only 8.3. "And the consumer can't measure that at home. You need a million-dollar lab to rate a unit's output. And we don't have a national, independent, random-sampling program to test whether companies are selling what they say they are selling. Compa-nies here can achieve an EER of 12, and they could sell a unit with such a rating for a 30 percent higher price. But there's nothing to stop a company from advertising and selling an EER-10 unit as an EER-12 and pocketing that premium."

Despite the doubts created by cheaters, political and economic pressures will surely bring true efficiency gains. But those gains can be totally eclipsed by the growth in air-conditioning use. A recent analysis found that a major move to higher-efficiency equipment would keep 119 million tons of carbon dioxide out of the atmosphere and the equivalent of \$3.7 billion in consumers' pockets annually when weighed against a scenario that assumes growth without increased efficiency; however, growing sales will reverse those gains, as a look back at Figure 3 makes clear. As a result, India's air conditioners will be sending eight times as much carbon into the air in 2020 as they did in 2005 despite predicted increases in efficiency.

HANDLING THE HEAT

Imperial Britain managed to rule India (including present-day Pakistan and Bangladesh) before the age of air-conditioning, but the subcontinent never saw very large numbers of pale Britons suffering the miseries described earlier by Reverend Merk. As Queen Victoria's reign was ending, there were barely twenty thousand British civilians sweating it out in a land whose population was surpassing 300 million (and many of the queen's representatives fled to cool "hill stations" in summer). Foreigners continued to avoid India's heat after the nation won independence in 1947. For almost half a century, most transnational corporations were shut out of India's economy by law, and Western tourists made sure to visit in the less hot winter months. Then, at the end of the twentieth century, the influx of Britons, Americans, and other heat-sensitive foreigners rose from a trickle to a steady stream, and many came as year-round residents. They were entering a country that remained, for the most part, at the temperature nature had assigned it; however, where comfort conferred the greatest economic benefit—airports, offices, computing facilities, upscale homes, hotels, cars—air-conditioning had already become well entrenched. In the economy of the 2000s, air-conditioned space has expanded in step with foreign involvement.

In India, the commonly used term "A/C" still translates as "VIP." Restaurants are divided into A/C and non-A/C sections, the former having pricier menus; an air conditioner can double a hotel room's nightly rate; air-conditioned taxis are much more costly; and a ticket for a berth in an air-conditioned sleeper car on the train system costs several-fold as much as a ticket in "Ordinary Class." When customers pay extra, they are not only covering the extra energy costs; they are buying a place apart from the rest of the population.

When the summer heat hits, most Indians adjust their daily routines, not a thermostat. In villages, sprawling suburbs, and central cities, people rely on a variety of low-tech ways to blunt the heat's effect. Stella, a woman in her early thirties who lives with her three young daughters in the rapidly growing town of Ramachandrapuram, a few miles west of Hyderabad, described the strategies that she, her family, and their neighbors put into effect in summer: "First of all, we like it better outdoors in the shade, because it's hotter inside. Children are on school vacation, so we all can stay out until eleven or twelve at night. When we go in, we'll sprinkle water on the floor to cool it and then sleep on a mat on the floor. We sometimes wet the curtains or even sleep under wet sheets. Or everyone will sleep outside; the men may all go up on the roof. We bathe a couple of times a day if there's enough water."

Stella says they generally eat their usual diet in summer ("We can't afford to buy things like fruit"), but they drink more fluids, especially ones reputed to help cool the body. For example, they take the water that's left after boiling rice, allow it to cool, and add some onion and chilies and a small bit of rice, and sometimes finger-millet malt as well. She and her family live in a sturdy concrete house and they have a ceiling fan. But the electricity always seems to go off when they need the fan most, and she uses it only sparingly in any case. She showed me her previous month's electricity bill, which, at seventy-four rupees ($1.50 at the exchange rate of the time, $4.50 in purchasing-power terms) represented a significant slice of her income. Running the fan more than she does would drive that cost up. When I asked what she'd do if she had an air conditioner, she laughed at the prospect: running it only twelve days in the summer would double her yearly electricity costs.

There are parallels between the America of forty to fifty years ago and the rural India of today. Recalling the fictional Mississippi sheriff's office in the 1967 film *In the Heat of the Night*—in which deputies hover in front of table fans, and the air conditioner in Sheriff Gillespie's inner office always seems to be on the blink—Gary Mormino told me that in the real Florida of the 1950s, if you put air-conditioning in your office, people would think that you considered yourself a big shot. In 1960, the *Florida Times-Union* asked the sheriff of largely rural Baker County about his new office air conditioner, and he said he'd decided it was now "safe" to install one. His use of the word "safe" indicates how negative the reaction would have been against an elected official indulging in such luxury.

In 2009, I encountered a similar situation in the small village of Gangadevipalli, which lies in cotton-growing country about a hundred miles east of Hyderabad. The village's sarpanch, or mayor, is Kusam Raja Mouli. He and his village have won numerous awards for improvements made in water supplies, sanitation, and the general economic climate. When, through a translator, he stressed the importance of responsible and transparent resource-management, I asked him what would be the consequences if he were to install an air conditioner in his office. Chuckling at the very idea, the sarpanch said, "First, people would say, 'This man's stock has really gone up!' Then they would say, 'He obviously is getting some illicit money from somewhere.' And by the way, there would be a lot of jealousy, on both counts! Of course, it would be used against me in the next election."

In urban India, officials are no longer so circumspect. Darshan Bhatia relates how he was once taken to a Hyderabad police station on a trumped-up traffic charge and held for an inordinately long time. He finally learned that he was being held because he had told the station's chief that he ran an air conditioner company, and the chief was hoping to be offered one of his products as a bribe.

"IF YOU CAN AFFORD IT, YOU ALMOST HAVE TO USE IT"

The McKinsey Global Institute reported in 2007 that with the entry of transnational corporations, the Indian air-conditioning market has seen intense competition. "This competition, combined with technological advances," according to McKinsey, "has improved the quality and energy efficiency of the products on offer. With market volumes and value estimated to have grown at over 20 percent annually over the past decade, this is a classic example of how the opening-up of a market has unleashed latent demand and boosted growth." India's "urban globals" (the 1.7 million people—only one-eighth of one percent of the nation's population—who, according to McKinsey, live in households with incomes of at least $118,000 per year in terms of comparative purchasing power) spend half of all money spent in India on appliances. Although the majority of households owning an air conditioner today are below urban globals on the income scale, the relative spending power of urban globals makes them the market to watch: "Apart from purchasing multiple items (e.g., several air-conditioners in a single household), these consumers will increasingly demand top-of-the-range, high-quality products similar to those used by the rich across the world." If, as predicted, large-appliance sales expand almost sevenfold between 2005 and 2025, McKinsey expects urban globals to continue accounting for half of the money spent, even though the class will still include less than one Indian in a hundred.

But spending in the upper reaches of the income scale is not done simply for the sake of spending. In 2009, the *New York Times* reported that makers of designer luxuries like $3,000 handbags and belts were having little success in expanding their market in thrifty India, even among people who could easily afford such extravagances. But truly practical luxuries like air-conditioning and automobiles continue to sell like ten-rupee masala dosas.

Research commissioned by the city of Mumbai has found that 40 percent of the metropolitan area's electricity goes to run air-conditioning. Discussing results of the 2009 study, the city power administrator could only say, "We are stunned." Looking to put the Indian air-conditioning market in context, I visited Girish Srinivasan in the offices of the Research Unit for Political Economy (RUPE) in Mumbai. The organization provides its analyses to grassroots groups across the country. I first asked Srinivasan if electric outages make cooling with air conditioners difficult in Mumbai. Although "there are long, sustained power cuts in the countryside each day, Mumbai is completely insulated from cuts that hit the rest of the state," he said. That reliable energy supply benefits "a small, wealthy sector that now treats air-conditioning as a necessity, and their electricity consumption is galloping ahead." To discourage excessive consumption, the government has set electricity rates to rise steeply at higher consumption levels, but "the rich are so rich, it doesn't affect their consumption. It's the same with petrol. Economic incentives don't work. To rein in luxury consumption, you have to impose physical restraints" such as resource rationing.

Srinivasan was speaking with me in late 2008, a couple of weeks after terror attacks had hit Mumbai and just as the Indian economy was being dragged down by the global financial collapse. Real estate was leading the crash. "During the luxury construction boom, we got shopping malls and 'automobilization,' both of which are heavily A/C-dependent. When the low-interest plug was pulled, the boom collapsed. But the rich can still easily pay for luxuries like A/C." He and RUPE founder Rajini Desai work in a cramped office in the heart of Mumbai's crowded, hectic Colaba district, close to several sites where the recent attacks had occurred. It's on a quiet side alley (or as quiet as it gets in Colaba), but Srinivasan said the pleasant December breeze I felt coming through the office's barred windows carried a heavy load of dangerous traffic fumes. That more than anything accounts for air-conditioning becoming a necessity for many in Mumbai. "As the common environment deteriorates," says Srinivasan, "there is very strong pressure to secure one's private environment." Thus, bottled water protects against intestinal illness and hepatitis, and air-conditioning protects against respiratory damage: "If you can afford it," he says, "you almost have to use it."

Air-conditioning is currently found in just 1 to 2 percent of Indian households. There is room for it to penetrate lower into the wealth scale, but at some point it will inevitably run into resistance, because the numbers of families who can afford it are limited. RUPE has published numbers demonstrating that the government and the media have exaggerated the size of India's growing middle class. Official sources tend to include a whole section of the population that doesn't have real middle-class consumption power. RUPE cites government figures from 2004–05 stating that about 230 million people, almost 20 percent, belong to India's middle class, based on the criterion that they spend an average of thirty-seven rupees per capita per day. That amounted to about eighty-three cents at the rupee-dollar exchange rate then prevailing; even with India's much lower cost of living, it would provide less than $2.50 worth of purchasing power—hardly what would be viewed in the West as middle class. RUPE estimates that the true middle class comprises 4 percent, not 20 percent of India's population. Along with the 2 to 3 percent who qualify as wealthy, they "constitute the booming consumer market that is the darling of global investors." That still amounts to almost eighty million people, a number not to be overlooked by marketers.

Yet Indian and multinational firms want more; they are looking to make inroads into that 90-plus percent of the population that lives below international middle-class standards. That requires low prices, easy credit, and, most important, big volume. "With these newer, less well-off customers, there is a much lower revenue per user," Srinivasan told me. "Corporations have tried very hard to make it pay, running airlines at a loss and selling cheaper cars and appliances. Television has penetrated into almost 20 percent of households. But the biggest success has been in extending mobile phones to a large segment of the population, across income levels."

Cell phone service is more accessible because it requires a smaller quantity of energy and resources per customer than do, say, air travel or personal cars. Air-conditioning, like planes and cars, is resource-intensive and therefore cannot be expected to diffuse through society as thoroughly as has mobile communication. Therefore, claims RUPE, international investors are eager to see wealth become more, not less, concentrated at the top of the pyramid, where buying power doesn't run into limits and there are plenty of rooms left to cool.

An economic development stampede would have hit India eventually, with or without air-conditioning. But the nature of that stampede would have been different without air-conditioning. Concentrated cooling power is essential to the highly visible industries that have led India into the international arena: information technology and communications. RUPE analysts argue that because the flood of foreign and domestic investment since the mid-1990s has been going almost exclusively into the corporate sector, and because that sector is becoming more capital-intensive with time, fewer new jobs are being created than is generally assumed—as few as four hundred thousand per year across the country. They write: "The vast majority of the workforce is trapped in sectors which are starved overall of investment: Agriculture accounts for the majority of the workforce and almost one-fifth of GDP, but accounted for only 6 percent of total investment in 2005–06 (down from 9.5 percent in 2001–02)." Jobs that are being created by the investment boom are mostly air-conditioned, white-collar jobs.

THE PRIVATE ENVIRONMENT

Each day, thousands of new personal cars join or displace the buses, trucks, scooters, autorickshaws, pushcarts, bicycles, and pedestrians that crowd the streets of India's large cities. By RUPE's estimate, private cars meet less than 5 percent of transportation needs but occupy 75 percent of the urban road space. Until recent years, few vehicles in India had electric turn signals or brake lights; therefore, hand signals were universally used throughout the twentieth century and are still common. Whether or not your car had working lights, driving safely in Hyderabad traffic as recently as the late 1990s meant keeping one's arm on the window ledge, ready to signal turns and stops; other drivers watched your arm, not your taillights. The rare air-conditioned car would display a warning message on its rear fender: "A/C Car—No Hand Signals." But today life has changed for the fortunate few. With air-conditioning, car owners are able to secure their "private environments"; a drive through traffic in any big city quickly shows that a large majority of personal cars now use their turn signals, their windows rolled up tight. Whatever the temperature outside, the pollution level is reliably high.

Commercial air-conditioning is spreading well beyond movie theaters, where it has been enjoyed by Indians of all socioeconomic classes since the 1970s. From 2002 to 2008, India went from having a grand total of six air-conditioned shopping malls covering a million square feet to more than two hundred and fifty air-conditioned malls totaling more than fifty million square feet. Construction of office space has exploded. New, air-conditioning-equipped luxury apartment buildings and single-family McMansions have mushroomed in and around major cities. In an economy that's scrambling toward the forefront of global capitalism, the environment is getting few breaks.

On city side streets, conservation is still the rule. Proprietors of small shops, just as they have done for decades, keep their low-wattage fluorescent "tube" lights turned off until a customer comes in. Most workplaces, including offices, continue to rely on natural ventilation; paperweights are a ubiquitous and essential feature of government and many private offices, where ceiling fans stir the warm air. But on Hyderabad's fashionable west side—the only territory seen by many foreign officials and investors—air conditioners work overtime in glass-clad, sun-battered office towers, and huge billboards remain brightly lit all night. Even the appalling American summertime habit of leaving a parked vehicle's engine running in order to cool the interior is catching on, despite high gasoline and diesel prices.

A manager in one of Hyderabad's top department stores, part of a national chain, told me that 55 percent of his hefty monthly electric bill can be attributed to air-conditioning. That is typical for freestanding stores in Hyderabad, he said, and it's close to the average for commercial buildings throughout India. Virtually all of the air is recirculated, because bringing in fresh but hot outside air would raise energy requirements to absurd levels. Cooling loads per square foot in the city's big new shopping malls are even higher, he says, because of the large crowds they draw, with each active shopper putting out a couple of hundred watts of heat. But human beings are really not so hot when compared with the growing ranks of computing devices that have led India's economic boom.

DIGITAL COOLING

The cool, green, glass-enclosed lobby of Microsoft India's Building Three just west of Hyderabad rises several stories into bright space overhead and connects two office towers, one with six floors, the other with eight. The only sound I could identify upon entering the lobby for the first time was the clack-clack of a table tennis game somewhere above my head. The atmosphere is relaxed and the amiable employees are dressed in bright tropical shirts and sandals, but there is nothing low-energy about this place. The Microsoft campus's enormous intellectual output is harnessed to a torrent of electric power. The campus lies in the stony, sunbaked country on the city's western outskirts, where triple-digit high temperatures can linger for weeks at a time, and summer temperatures routinely top 110°. In the campus's three buildings, with floor space totaling 1.3 million square feet, software engineers work in upstairs offices on computers connected to countless racks of servers in four ground-floor "laboratories." Computer chips are built to withstand high temperatures, but they process data better when they are cooler. It falls to Keith Dias, senior manager in charge of facilities, to ensure that neither the employees nor the servers overheat, no matter how scorching the weather outside. Offices are set for 75°, says Dias, with server labs usually just under 72°. In summer, that can mean reducing temperatures by 40° or more below those felt outdoors in the shadow of the buildings.

The Microsoft Hyderabad power demand comes to 13 megawatts, enough to supply perhaps ten thousand typical U.S. households, averaged through the year (or almost 140,000 Indian households). About 40 percent of that goes to run the servers, computers, and related equipment, and another 40 percent runs the air-conditioning systems. In data centers, it's typical for the electric bills racked up by computing and cooling equipment to be approximately equal; that's not surprising, since the bulk of the wattage consumed by a computer is converted to heat. In India, software firms, like farmers, are key economic players who take their share of the nation's power supply right off the top. Dias, therefore, has few worries about electricity supply: "In the five years I've been here, we've never been subjected to load-shedding. When sporadic outages occur, we are fully backed up with diesel generators." But, he says, backup is expensive: "Energy from the generators costs one and a half times as much as that from the utility."

On a pleasant winter day, 80° and sunny outside, Dias and I made our way into the bowels of Building Three, entering one of four dim, slightly chilly rooms where rack after rack of servers stand, strategically arranged so that conditioned air can be placed right where the devices' fans can draw it across circuits blazing with data. The machines, many of them "blade servers" whose computing power and resulting heat are packed extremely densely, are clustered according to the specific software-development group within Microsoft that uses them. As we entered the labs from the hallway, I felt a distinct drop in temperature. Indeed, on that day, the air in the server lab was hovering around 68°, well below the temperature Dias prescribes. He instructed the lab's manager to move the thermostat up and told me, "Some of the guys push us to get it down as low as nineteen [66°F] in the labs. And some insist on keeping server racks up in their offices, because they want to live and sleep with them. But offices aren't built to handle the extra heat. We also have no set office hours here, and some employees like to work at any hour of the day or night; that means running the A/C system all night just for a couple of people." All of that, he says, makes efficient climate control more difficult.

Because employees' brains also work most efficiently at comfortable temperatures, says Dias, a software firm's indoor climate is crucial to its output: "Our working conditions need to be peerless, because we have to attract the very best people here." As an added incentive, Microsoft is one of a tiny handful of area companies that provides air-conditioned buses and vans to transport all workers to and from their homes.

THE GREEN BUSINESS

In the final chapter, we will examine how, in India and other hot regions, comfort can be maintained without refrigerated air. There are also small but growing efforts to use air-conditioning more efficiently in the workplace, as part of a larger effort to soften the environmental impact of rapid economic growth.

S. Srinivas, principal counselor with the Confederation of Indian Industry (CII) in Hyderabad, goes to work each day in India's greenest building. He knows precisely how efficient and ecofriendly the building is, because he helped design it. With one-thirtieth the floor space of Microsoft's Building Three, the CII-Sohrabji Godrej Green Business Centre is barely visible from the main road, its intriguing arcs and angles tucked into a rolling, five-acre patch of green in the concrete jumble of Hyderabad's Hi-Tec City. (The immense glass cliff looming just across the street is the south face of a Google facility.) The Green Business Centre was the first building outside the United States to be certified under the Leadership in Energy and Environmental Design (LEED) rating system as "Platinum," and it sports a full array of energy-saving features, green construction materials, and its own water-treatment garden to completely recycle the water used by the building. The biggest challenge of all, says Srinivas, is to fend off Hyderabad's aggressive heat. To that end, the external walls are built of fly-ash blocks, which transmit only a third as much heat as does concrete. All windows are designed to minimize solar heating of the interior. "These windows insulate like clay bricks," says Srinivas. The roof is composed of similarly heat-tight extruded polystyrene, and two-thirds of it is covered with a thin layer of soil sown to grass. When the grass is watered, it transpires water vapor into the hot, dry air, providing an extra cooling boost. Another large area on the roof's south slope is covered by a photovoltaic array supplying 24 kilowatts. In addition, Srinivas says, "all fans and pumps have variable-speed drives, and all appliances are the most efficient available."

The Centre's roofline is merrily unconventional, but its footprint is in the traditional Indian courtyard shape; both features help light to penetrate deep into the building. Several other minicourtyards throughout the building admit additional light. On the afternoon that I visited, I saw no electric lights burning anywhere; that in itself makes cooling easier. But by far the building's most highly visible energy-saving features are two wind towers, based on technology developed in Persia centuries ago. The slots and spaces in the towers' concrete are wetted each morning—with the building's own treated waste water—and air is drawn through before entering the conventional air-conditioning system. "That precools the air as much as 10° to 12°C [18° to 22°F] and saves 7 to 8 percent on total electricity consumption," Srinivas told me. The towers, which provide their biggest benefits during that three-fourths of the year when humidity is low, are doubly important because the building's designers put an almost obsessive emphasis on fresh air. To prevent sick building syndrome (Srinivas says, "The air in the new shopping malls will choke you!"), carbon dioxide sensors signal the system to flush the building with large volumes of outside air when needed. The wind towers relieve some of the burden that all that overheated outside air puts on the air-conditioning system. The Centre's thermostat is kept at 77°, but on a low-humidity day it seemed cooler than that. Maybe the building's architectural wonders were working the power of suggestion on me, but the general atmosphere, physical and otherwise, seemed far superior to that of a normal office building.

The Green Business Centre's bottom line is an average draw on the local electric utility of less than one watt per square foot, compared with typical office building intensities of four to ten; data centers or server labs like the ones on the Microsoft India campus have intensities ranging up to 200 watts per square foot. Work is done in the Centre as in any office building, but its chief function is simply to be green and serve as a model. Is the model being emulated?

Hariharan Chandrashekar says it is. As CEO of Biodiversity Conservation India Limited (BCIL), headquartered 300 miles south of Hyderabad in the equally fast-growing high-tech city of Bangalore, Chandrashekar opposes pumping energy into more urban glass towers. During the boom of the early 2000s, he says, "people weren't conscious at all of energy consumption. They'd say, 'We'll build as much as we have the money for.' For a model, they would take a building they saw in Illinois or somewhere." Now, says, Chandrashekar, things are changing. "We've set up very straightforward green guidelines that any architect, builder, or electrician can understand. By 2012, we expect that one billion out of six billion square feet of office floor space will be green." The new space may not be as thoroughly green as the Green Business Centre, but he expects to be much more energy efficient.

In practice, the green growth projected by Chandrashekar will mean adding many millions of square feet of air-conditioned office space to an economy that once ran solely on natural ventilation and fans. Meanwhile, housing and work environments in the country's less well-off areas will remain uncomfortable by Western standards. Once again, conservation will be measured as the difference between efficient but rising resource consumption by a small, elite segment of the economy and a worst-case "business-as-usual" scenario.

8

INCONSPICUOUS CONSUMPTION

You've got your ice. You've got your heat. Sell the eggs you don't need and have the rest for breakfast. Cool down your vegetables. Use your chicken shit for methane. It's a perpetual-motion machine. Run a duct to your house and you're air-conditioned—cool in summer, warm in winter. Cheap, simple to operate, no waste, foolproof, and profitable.

—Paul Theroux, *The Mosquito Coast*, 1982

The chief source of problems is solutions.

—Eric Sevareid, 1970

What if the many trillions of dollars of annual output from the worldwide economy did not represent any real production? What if it were not true output but rather what ecologists call "throughput," and what if, when everything is accounted for, the bulk of matter and energy coming into the economy were fully accounted for by the irretrievable waste matter and energy coming out, leaving no tangible product? Vanderbilt University economics professor Nicholas Georgescu-Roegen arrived at that very conclusion in the 1960s. Relying on the formulas of classical economics, he showed that an economy's only product—indeed the ultimate goal of all economic activity—is the nonmaterial "enjoyment of life," which has no economic value and can be accumulated only as memories. To produce enjoyment of life, the economy's material flow is continuously converting useful matter and energy into wastes that cannot be reused, in accord with the laws of thermodynamics; the second, or entropy, law dictates the irreversibility of the process. It sounds grim, but there's really no major problem. If his model holds, humanity still can achieve a good quality of life. But the more we can reduce throughput, the longer that good life can continue.

Georgescu-Roegen is considered the philosophical father of the discipline now called ecological economics, and air-conditioning provides one of the clearest illustrations of his worldview. Physical comfort, most of us would agree, is an important part of the enjoyment of life. And the throughput of energy and matter—from the mine, the well, and the ecosphere through the air-conditioning system to the atmosphere and waste dump—is well understood. The air conditioner depends on the second law of thermodynamics for its operation, and the moment the device is turned off, the fruits of its labor start to be undone by that same law. The ephemeral enjoyment of life it has produced is very real and very welcome. The material throughput that the refrigeration process requires to maintain that enjoyment can be reduced, but not beyond a certain point and never close to zero.

Despite that, many "green"-economy enthusiasts go about their work as if the second law had been repealed. Capitalist economies run on the assumption that any serious interruption in the accumulation of material wealth—as happened in a spectacular way in the 1930s and again in recent years—is always temporary, and long-term growth cannot be restrained. This line of thought was shared by the protagonist of Paul Theroux's novel *The Mosquito Coast*, for whom manufacturing ice in a Central American village is "the beginning of perfection in an imperfect world." Impatient with the boundaries set by physical laws, Allie Fox lectures his fellow villagers, "Why do things get weaker and worse? . . . Why don't they get better? Because we accept that they fall apart! But they don't have to—they could last forever." Unfortunately, there are no perpetual motion machines for making ice or anything else, and we cannot prevent things from "getting worse" overall. But we can ensure that they don't get worse at a catastrophic rate.

Before considering new approaches to air-conditioning, natural cooling, and thermal comfort, it is first necessary to wrangle with some current approaches that won't work—ones that could well create more problems than they solve.

DEMATERIALIZATION

A few economists have looked at the current situation with clear eyes, and without invoking perpetual-motion machines. According to a 2008 analysis by Minqi Li of the University of Utah, the world economy must contract at a historically rapid rate if atmospheric carbon dioxide is to be held below 445 parts per million (ppm), a level not even predicted to keep the world the way it is, but a cap meant merely to fend off the most calamitous climate scenarios. Because a substantial proportion of the world's population is already living at or below subsistence levels, the reductions must come almost entirely in better-off nations. His range of negative economic growth estimates—shrinkage of the total world economy at an annual rate of -1 to -3.4 percent through 2050—is based on a range of energy scenarios going from business-as-usual to the most dramatic of feasible improvements in efficiency and alternative technologies. In all of his scenarios, however rosy their assumptions, economic growth will have to be thrown into reverse, or else. Reducing atmospheric carbon dioxide to 350 ppm and holding it there—as is advocated by the international movement "350.org," co-founded by writer and environmentalist Bill McKibben—would be expected to require much deeper cuts in growth under Li's model. Such reversals would be incompatible with the functioning of our capitalist economy and will not be undertaken voluntarily by national and international decision makers.

The U.S. Department of Energy reported that economic recession reduced U.S. carbon emissions by 2.8 percent in 2008 and would bring a further 5 percent decline in 2009. In previous decades, the United States' carbon emissions grew steadily, the only exceptions coming in other recession years: 1981–83, 1990–91, and 2001. Reducing economic output is a tried-and-true way to reduce emissions, but when the shrinkage is unplanned and chaotic, as in all past recessions, it creates more misery than anything else.

In the absence of proposals for orderly, equitable shrinkage, economists and ecologists in various shades of green continue to argue for "dematerialization"—the generation of higher economic output per unit of resource input. That would, it is argued, permit wealth and incomes to rise indefinitely while resource throughput and ecological destruction are reduced. It is to be achieved by converting from manufacturing to a service economy and by altering the nature and content of products and services to take advantage of human innovation and use smaller quantities of nonrenewable natural resources. Economist Herman Daly compares the economic models that promote dematerialization to a recipe by which "we could produce a 100-pound cake with only a pound of sugar, flour, eggs, etc., if only we had enough cooks stirring hard in big pans and baking in a big enough oven!"

In a dematerialized world, nonmaterial goods and services are supposed to displace material ones. Such an economy may rely upon more and more rapid circulation of words and images, advice and conversation, messages and massages. But can a nation, let alone the world, satisfy itself by replacing material goods with ghostly data and personal services? How much will consumers be willing to spend with nothing tangible to show for it? And if dematerialization works, where will the growing stores of wealth end up? Don't they eventually and inevitably seek out something concrete? Both the theory and experience of capitalism say they do.

Once again, an excellent case in point involves air-conditioning in the leading role. The fast-growing digital economy, regarded as a crucial engine of dematerialization, is in reality a voracious gobbler of material resources, including cooling power. The EPA estimated in 2006 that network data centers—large computing facilities that do much of the Internet's work, along with processing bank and credit-card transactions—had doubled their energy consumption in just six years. The server labs on the ground floor of Microsoft's Hyderabad campus are big but look like puny closets compared to the vast "server farms" that keep world networks going. Take the energy consumed by today's data centers, throw in the growth projected for the near future, along with the energy used by computers and peripheral equipment in homes and commercial buildings, and by 2011 the United States' computing network will require a quantity of energy annually equivalent to Mexico's total electricity consumption, or, by 2014, Australia's. The EPA estimated that by 2011 the peak load placed on the grid by servers and data centers alone would require the output of twenty-five normal-sized power plants.

Each generation of technology, from the vacuum tube to today's advanced processors, has handled more information per watt of energy input, but that efficiency has always been harnessed to push speed and output higher, not to conserve energy. The industry has produced tinier chips and bigger, denser, hotter arrays of chips every year. The lion's share of the wattage going into a data center's computing equipment comes out again as waste heat, so the total wattage consumed by computing is included when calculating the load on air-conditioning systems. Running and cooling a single six-foot-high rack of servers occupying seven square feet of floor space can consume as much power as would thirty typical California homes. Thousands of such racks in rooms or buildings ranging into the hundreds of thousands of square feet can impose enormous cooling demands.

In line with 1960s-era predictions, the computational power of a given size of silicon chip continues to double every eighteen months. That and other improvements have allowed the computing performance of servers to triple every two years. Without improvements in energy efficiency—in the amount of information servers can process for a given input of electricity—the system would have melted down long ago from the heat generated by denser machines and the resulting growth of utility bills. Efficiency, out of necessity, has been greatly improved, but it has only doubled, not tripled, every two years, thereby failing to keep up with energy consumption for running and cooling servers. It's not even close. In the brief period between 2000 and 2006, the amount of energy to process a given amount of information fell 88 percent, yet energy consumption for a given investment in data handling rose 300 percent. And the numbers and sizes of data centers have swelled.

The industry continues to invest in efforts to reduce heat output per quantity of information processed and to improve cooling technology. Where the limits of air cooling are approached, liquid cooling may become common. With the past and economic reality as guides, we can assume that an improved capacity to deal with heat stress will be used to expand computing power further rather than reduce the industry's energy footprint.

Internet exponents claim that ever-bigger expenditures of energy are compensated for by the many resource efficiency gains that computers make possible. But such gains, where they have occurred, are overwhelmed by our general resource use. Where progress has been seen, economic contraction, not technology, was chiefly responsible:

- Electronic communications were expected to cut into paper use, but savings have been slow in coming. Paper consumption for all uses in the United States had hit a peak of more than 700 pounds per person annually by the 1990s, a 25 percent increase over the pre-e-mail days of the 1970s. A 3 percent drop in paper use in the 2000s may mean that computerization is finally having a small impact. But office printing technology gets better every year. Even if printed books, newspapers, and magazines were to disappear completely, would business in general really use decreasing quantities of paper in prosperous times?

- Online shopping was supposed to help limit the size of the climate-controlled, brick-and-mortar retail world and keep shoppers out of their air-conditioned cars. As we have seen, the reverse happened; square footage of store parking lots and climate-controlled retail space per person continued to rise right up until the 2008 bust.

- It is widely anticipated that videoconferencing and telecommuting will substitute increasingly for business travel. But industry data show that, aside from a short post-9/11 slump, U.S. business travel marched upward at a steady rate of 5 percent per year from 1990 through early 2008. Only the economy's plunge managed to stifle the urge for business-related flying and driving.

Overshadowing all of those issues, perhaps, is the Internet's capacity to stimulate more consumption than it eliminates. A big share of that energy going into running and cooling data centers is aimed at convincing Web users to consume more of everything, to convert more matter and energy into waste somewhere else. For example, revenues from Internet advertising almost quadrupled just between 2002 and

2008. Another big share of the Internet's energy goes to processing the orders and payments stimulated by the ads; 25 percent of all traffic through search engines goes to retail sites.

THE MIRAGE OF EFFICIENCY

Improving the energy efficiency of air conditioners and other devices is a widely discussed route to lower greenhouse emissions and reduced resource destruction. But benefits from efficiency can be elusive. For example, the U.S. Energy Information Administration reported in 2009 that improved efficiency in all kinds of electrical appliances is being nullified by increased air-conditioning use:

> Residential electricity use has increased by 23 percent over the past decade, as efficiency improvements have been more than offset by increases in air-conditioning use and the introduction of new applications. That trend continues. . . . In 2030, electricity use for home cooling in the reference case is 24 percent higher than the 2007 level, as the U.S. population continues to migrate to the South and West, and older homes are converted from room air-conditioning to central air-conditioning.

Despite such setbacks, some environmentalists continue to predict positive ripple effects from technological efficiency. For instance, Hariharan Chandrashekar, the green-construction advocate in Bangalore, told me, "If people save money through efficiency, they can afford further green improvements, which will save even more money. You have to get into that virtuous circle." But environmental scientists Mario Giampetro and Kozo Mayumi have argued that economies, as "complex adaptive systems," are unlikely places for such virtuous circles to be generated, and that they defy analysis in terms of simple efficiency. The definitions of products being sold, bought, and used can change dramatically even over short periods, not only making it hard even to define "efficiency," but also giving economies a tremendous capacity to take advantage of efficiency improvements through expansion. Often those improvements are turned on their heads. Giampietro and Mayumi illustrate:

> Looking at the evolution of cars in time, we can say that the introduction of more efficient car engines has determined that some features, such as air-conditioning, which were optional in the past became standard features of modern cars. Thus an increase in efficiency in one of the attributes of performance . . . has led to the addition of a new set of standard attributes in the definition of "what modern cars are and should be."

They observe that when any technical improvement in resource efficiency is achieved, a society has to choose how to take advantage of each increment of improvement: either to use a smaller quantity of resources (which includes putting a smaller burden on ecosystems) while maintaining the current material standard of living, or to increase the material standard of living at equal or higher resource use. They conclude that this question must be resolved through the political process. If an explicit political decision is not made, the latter option—increasing per capita production and consumption—will be adopted by default in a capitalist economy because of the imperatives of growth and wealth accumulation. That, of course, leads to expanding resource use.

Giampietro and Mayumi also observe that "to increase efficiency now, one has to eliminate obsolete solutions from the existing portfolio." In a world of limited resources, adopting a newer, more efficient technology can mean eliminating older solutions that might be needed later. Our cities, for example, are full of tight, energy-efficient buildings designed on the assumption of air-conditioning. They would be far less habitable than the "inefficient" buildings of earlier eras if natural ventilation were used instead.

A car's fuel efficiency is easily tracked using its odometer and the meters on gas pumps. The efficiency of cooling cars or buildings is not so simply estimated. Recall that in India, there are suspicions that energy-efficiency claims for air conditioners are being fudged, and even the most accurate estimates can be misleading. In the late 1990s, Mithra Moezzi of the Lawrence Berkeley National Laboratory summarized what he called "the predicament of efficiency" for household appliances. The problem, he wrote, is that when agencies confer efficiency rankings and awards, as the EPA does under its "Energy Star" program, they use the standard definition of high efficiency: more service delivered per unit of energy consumed. Products are ranked only against others of similar size with similar features, so appliances that use far more energy than more modest models are often declared, perversely, to be the most efficient. Thus, Moezzi noted, "an electric toothbrush may be labeled as efficient while a manual toothbrush will not be." He cited a "Golden Carrot" award bestowed by the Consortium for Energy Efficiency on a Whirlpool twenty-two-cubic-foot side-by-side refrigerator that was more highly "efficient," but only when compared with other side-by-sides. But the side-by-side design (with the freezer compartment on one side and the refrigeration compartment on the other) is inherently inefficient compared with the traditional design, and large refrigerators have inherently higher energy requirements. Thus, the award-winning model's electricity consumption exceeded the maximum allowed for similarly sized freezer-on-top models under national standards. Moezzi also cited the 1997 "Your Energy Star Home" calendar, which came out during the rapid acceleration of U.S. house-size growth. Of the seven energy-efficient houses featured, four were larger than 3,800 square feet. As we have seen, any house of that size will use far more energy than an "inefficient" 1,500-square-foot house.

The problem persists. In 2008, it was revealed that two refrigerators with "French door" cooling compartments and freezers on the bottom—one made by Samsung, another by LG—qualified for Energy Star ratings only because they were allowed to be tested with their automatic ice makers turned off. Thus tested, they each consumed at rates of under 550 kilowatt hours per year. But with the ice makers on, the Samsung used 890 and the LG a whopping 1,100 kilowatt hours per year. Even with the ice makers off, both Energy Star models were more power hungry than some of their somewhat smaller but still capacious contemporaries that did not qualify as "efficient," such as a twenty-two-cubic-foot, freezer-on-top Maytag that had an annual energy use of only 448 kilowatt hours. Clearly, efficiency remains too ill-defined to be of much use in the consumer marketplace. From such contradictions, Moezzi concluded,

> One of the difficulties in moralizing about energy consumption is that most energy, and thus most energy savings, is invisible at the point of use. Using labels as rewards for increased efficiency and as a means to convey information about "high-efficiency" systems is an attempt to make energy use and savings visible. This is a good strategy if the goal is to get customers to buy, which is a foundation of a market-driven approach to energy efficiency. However, there may be some long-term danger in awarding ratings that imply environmental beneficence to activities that could hardly be considered environmentally beneficent.... Labels implying that energy efficiency leads to conservation are misleading if they cause people to buy more, or larger, products than they otherwise would have.

Luring customers, in effect, to purchase an energy-efficiency label in lieu of actual efficiency could turn out to be an even more clever moneymaking strategy today than it was in the 1990s when Moezzi wrote that. With society now placing increased emphasis on a green lifestyle, well-off people may buy seemingly efficient refrigerators or air conditioners not only out of a desire to save on future energy bills but also to strengthen the buyer's reputation as a responsible citizen—a green version of conspicuous consumption. On the other hand, Lancaster University professor of sociology Elizabeth Shove views the routine use of air-conditioning in Western societies today as an example of "inconspicuous consumption"—done for what is thought to be a necessary, practical purpose rather than simply to gain status. Therefore, she argues, decision makers like to focus on efficiency instead: "In concentrating on *efficiency* rather than consumption, policy makers stick close to a politically safe position, providing information and advice but not going so far as to tell consumers and decision makers how to live their lives. ... In effect, demand—including demand for air-conditioning—is taken for granted and so taken out of the equation." So consumption is free to rise right along with efficiency.

Manufacturers have made gains in air conditioner efficiency over the past thirty years. Residential central air-conditioning units in service in 2005 were an impressive 28 percent more efficient on average than those in service in 1993, based on their SEER (seasonal energy efficiency ratio). But the average household's air-conditioning system used 37 percent more energy in 2005 than in 1993. (And because increasing numbers of houses were being air-conditioned, total energy use rose to a level double that of 1993. See Table 3.) Much-improved, more energy-efficient equipment was consuming a lot more energy per home. Federal standards were tightened in 2006, requiring that new equipment be another 30 percent more efficient. Should we expect another 30 percent increase in energy use as a consequence?

Economists have labored heroically over the past quarter-century to accumulate evidence for or against this efficiency/consumption enigma, one that had been articulated in 1865 by economist William Stanley Jevons in his book *The Coal Question*. Having seen that major improvements in efficiency of coal-fired steam engines over the previous half-century had been accompanied by a rapid expansion of coal consumption, Jevons argued that improved efficiency of resource use will inevitably result in increased, not decreased, consumption. The seeming paradox came about, he wrote, because more efficiently produced goods can be produced more cheaply, which in turn increases both effective production capacity and demand, stimulating the entire economy.

Following the energy crisis of the 1970s and accelerating into the 2000s, economists started peering more deeply into Jevons's paradox, both theoretically and through analysis of real data. As a result, his simple proposition has been stretched into a more complex continuum featuring the concepts of "rebound" and "backfire." Suppose that technical improvements in air-conditioning mechanisms and/or better insulation allow home owners to expend less electricity in cooling their houses. If they respend some of the savings that show up on their utility bills by turning their thermostat temperatures lower or running other appliances more often, total electricity consumption won't fall as far as is suggested by the degree of efficiency improvement. That is called a rebound effect. If consumers really get carried away and end up using just as much electricity as they did before the efficiency improvements, 100 percent rebound has occurred. And if, as the Jevons paradox predicts, rebound surpasses 100 percent—that is, if total electricity consumption actually increases—that's backfire.

We saw an example of apparent backfire in the previous section: fast-rising power consumption to run and cool ever-more-efficient computer-network servers. If air-conditioning is viewed strictly as a consumer good, it's not hard to imagine energy savings from more efficient manufacturing, greater operating efficiency, or more weathertight buildings being at least partly canceled out by rebound. In India, where the market is far from saturated and electricity costs are high compared with the original investment in cooling equipment, more energy-efficient models could lead to more widespread adoption of A/C, and owners may decide to operate units for more hours per day. In the largely saturated U.S. market, high cooling efficiency could offset electricity price hikes and keep electricity consumption high when switching off appliances, opening windows, and using attic fans might otherwise have started looking more attractive. And because efficient climate control can be a powerful economic stimulus package, it could spur consumption throughout society at large.

Government programs for insulating low-income homes in hot climates have run into strong rebound effects. A study published by the Oak Ridge National Laboratory in 2008 found that energy use for air-conditioning in south Texas homes was not significantly reduced through weatherization, partly because residents took advantage of the improvements by keeping their houses about two degrees cooler, on average, than they had done previously. Similarly, an experiment conducted by the Florida Power and Light Company in the early 1980s measured the effects of three energy-saving technologies that the company provided free of charge to randomly chosen consumers: insulation, insulation plus a more efficient air conditioner, or insulation plus an efficient heat pump. A fourth group of customers received no improvements. The results showed that although "engineering models

often assume that a given percentage improvement in thermal efficiency . . . will translate into an identical percentage reduction in electricity usage," that was not the case, because "home owners will use
their air-conditioning and heating more intensively when the effective
price of comfort is lower."

Rebound or backfire can be either proved or disproved theoretically, depending on the economic assumptions and models you use, on
whether you stick to a single product or resource, and on whether you
consider the entire economy or just a sector within it. Estimates based
on actual data, usually limited to individual products and inputs, have
varied widely as well. You can find energy-rebound estimates of 0 to 50
percent for air-conditioning, 10 to 30 percent for heating, 10 to 40 percent for water heating, 5 to 12 percent for lighting, 65 percent for overall home electricity use, 5 to 25 percent for home weatherization, and
5 to 50 percent for vehicle fuel consumption. Research is sparse, but the
mechanisms through which rebound can occur are easily imagined.
For example, how do people respond to advice they are given on fluorescent bulbs: to use them primarily in fixtures that are turned on for
several hours at a stretch and to leave them burning at least fifteen
minutes before turning them off, all to prolong their life? Do we tend
to keep them turned on longer each day than we do incandescent bulbs,
thereby eating into advertised energy savings? And to what degree can
green efficiency with one resource lead us to consume more of another? For example, do people with solar water heaters tend to take
longer showers? Will the advent of ozone-friendly, nongreenhouse refrigerants lead some ecoconscious consumers to install a new air-
conditioning system rather than considering less energy-intensive
alternatives?

For none of those separate slices of the economy does 100-percent
rebound or backfire appear to occur. Yet when modern economies are
observed in their entirety, efficiency gains and rates of resource use almost always seem to march upward hand in hand, just as they did in
Jevons's day. That, say backfire believers, is because efficiency saves
money and thereby expands the general power of firms to produce and
of people to consume. The entire system grows in response. Because
our economy is designed to never leave a usable resource unused,

higher consumption is inevitable. Economies as they operate today always behave as if Jevons's paradox is true.

Over the years, mainstream economists have often denied the existence of the paradox despite real-world evidence of rebound and backfire. But the question of whether or not efficiency "causes" the increased consumption is best set aside as a subject of debate for university seminars. Resource efficiency on the scale of a household or company almost always benefits those who put it into practice, but that does not automatically conserve resources or protect ecosystems across a city, state, nation, or planet. Leonard Brookes, a British economist who shares credit for the modern incarnation of Jevons's paradox—known as the Khazzoom-Brookes postulate—views energy efficiency as a distraction: "Maximizing energy efficiency has no particular merit as a national or international target. It is not a proxy for maximizing economic efficiency or social benefit or minimizing environmental damage. Pursuing it as a target entails bias that in turn leads to misallocation of available economic resources. Claims made for progress in pursuing it are often faulty."

Given no politically palatable alternative to growth, elected officials tend to gravitate toward energy-saving measures that require little or no sacrifice. That, of course, is also the kind of legislation that is least likely to save energy. In 2005, the U.S. Congress passed a law extending daylight saving time (DST) by three weeks in the spring and one week in the fall, effective in 2007. (The idea behind DST goes back to Benjamin Franklin, who suggested moving the clock ahead in summer to reduce the amount of time that home lighting would be necessary during evening hours, thereby saving vast quantities of tallow and wax.) But in 2008, researchers at the National Bureau of Economic Research spoiled the fun when they reported their results from a three-year study in Indiana. (The state offered what they called a "natural experiment" because it had switched to statewide DST only in 2006; previously, only a few of its counties had followed DST.) They drew these conclusions:

> In all months other than October, DST saves on electricity used for lighting. . . . But when it comes to cooling and heating, the

clear pattern is that DST causes an increase in electricity consumption. The changes in average daily consumption are far greater for cooling, which follows because air-conditioning tends to draw more electricity and DST occurs during the hotter months of the year.

The bottom line of the switch to DST: a 1 percent increase in electrical consumption in Indiana, at an annual cost of $9 million in higher residential utility bills and $1.7 to 5.5 million in expected costs to be incurred because of the resulting pollution.

CLEAN-ENERGY DELUSIONS

With painfully little time in which to turn the tide on ecological ruin, both the conservation measures and the energy technologies on which we place our bets now had better be the right ones. If they aren't, we may not have time to switch later. If we plan to convert to renewably generated electricity in order to keep indoor America lighted, cooled, and otherwise powered up, we have a long way to go. As the numbers in chapter 2 show, covering the many crucial uses of electricity in a sustainable way will involve hard choices. Even with crash programs to install energy-efficient lighting and good insulation, it's unlikely that truly renewable capacity or carbon sequestration can be expanded as quickly as fossil-fuel-based capacity needs to be cut back. If greenhouse gases are to be controlled, very deep cuts in electricity consumption will be required.

Political and economic decision makers don't like forcing choices like that, so it's usually at this point in the discussion that the biggest deux ex machina of all, nuclear energy, is rolled out. Public support for a "nuclear renaissance" is being pumped up by the erroneous impression that nuclear plants produce electricity with few or no carbon emissions. Policy makers and even some environmental organizations and environmentalists who should know better have been only too happy to step to the front of the chorus.

Nuclear plants currently account for about 10 percent of the nation's electric generating capacity and satisfy about 20 percent of total consumption (because they run closer to capacity a greater portion of the time than do other types of power plants). If we are to rely on nukes to replace coal and other fossil fuels in satisfying demand for electricity, we'll need to start building new plants again, and fast. Look at the nation's peak electricity demand in summer, which exceeds the winter peak by 144 billion kilowatt-hours. To cover that additional peak demand—most of which can be chalked up to air-conditioning and refrigeration—would require the collective capacity of all 103 existing nuclear reactors in the United States, plus the construction of forty-three additional average-size reactors. That would leave no nuclear-generated peak power for anything else, and to cover increased cooling demand, additional plants would have to come on line every year.

In 2008, Joshua Pearce, assistant professor of mechanical and materials engineering at Queen's University, published figures showing that if we were to cut world greenhouse emissions by 60 percent in an effort just to "stall" global warming, and if all non-fossil-fuel energy were to be supplied by nuclear plants, the number of nuclear plants worldwide would have to increase from the current 350 or so to almost 8,000. To cover projected growth in world energy demand up to 2050 with nuclear energy would take another 18,500 plants; in total, then, we would be putting into service 1.8 nuclear plants *per day* over a forty-year period. That's clearly not in the cards, but suppose we did launch a do-or-die effort to save the planet by building nuclear capacity as fast as possible. Any such effort will have to count on the negative side of the ledger all greenhouse emissions created in the process of building nuclear capacity.

Nuclear generation is "carbon-free" only if you ignore the resources consumed in constructing reactor plants, mining and refining uranium and other minerals, transporting materials, handling wastes, and decommissioning the plant. Taking all of that into account, nuclear plants still do have far lower greenhouse potential over their lifetimes than do coal- or gas-fired power plants. But the enormous amounts of energy that must be spent in advance to create the capacity required to generate nuclear energy in the future would require us first to go into deficit carbon spending for years. Pearce examined the energy and emissions costs that would be incurred with a rapid buildup in America's nuclear capacity, taking into account the entire life cycle of nuclear plants and using the most complete data available on the energy requirements of each phase of that cycle. He concluded that with a high nuclear-industry growth rate, carbon savings provided by newly built plants will be "cannibalized" by the heavy carbon costs of building more plants each year and dealing with ever-mounting wastes. Therefore, during the very period between now and 2050 when deep cuts are needed to prevent runaway warming, we would be nullifying our progress. In the all-nuclear scenario, *all* emissions reductions achieved up to 2050 would be canceled out by growing emissions from plant construction, supply, and operation. Capturing the vast quantities of waste heat coming out of plants and using it to produce additional electricity (which is now done at some plants in Europe but not in the United States) would brighten the picture. Pearce says the full life-cycle costs of doing that have not been calculated, but when they are, he thinks the results will show that it will be much less costly to install waste-heat exchangers than it will to build additional reactors.

"Breeder reactors," which produce new fissionable material faster than they consume the material they're fed, recur as a potential source of abundant fuel, but the processes required to produce fuel for and run them are extremely dirty, dangerous, and expensive. Says Pearce, "Breeder reactors have a long list of contamination problems. To make a dent in climate change, it would have to be done on a really big scale. I'm not sure anyone would be organizing for that on a large scale. That would be crazy." He sees a need to spend our carbon wisely now to build an infrastructure that can save carbon later: "Solar energy grew at close to a 50 percent rate last year; that made it a net carbon *emitter* for the year! No problem; we can overshoot a little. That's OK." But, he says, we can't hit that atmospheric concentration that represents a point of no return: "We have to get all this stuff on line before we reach that threshold." Solar, wind, geothermal, waste heat—all, he says, have more potential to reduce emissions in the near future than does nuclear energy.

The hottest issue in the nuclear industry continues to be the handling of its waste products. The question of a permanent home for

62,000 metric tons of spent nuclear fuel rods now stored alongside the reactors from which they came has so far stumped America's scientific and political leaders. Bomb making, power generation, and other uses of radioactive elements leave behind a wide variety of unneeded, hazardous isotopes. One of them, cesium-137 (^{137}C), is what makes spent fuel rods fresh out of the reactor extremely "hot" (in both the thermal and the radioactive sense) and dangerous. Today in America, fuel rods are almost always stored on-site as they come out of the reactor, in nearby pools of water; a typical pool holds four hundred tons of such rods, five times the tonnage in the reactor core itself, with twice as much ^{137}C per ton of rod. A study led by Robert Alvarez of the Institute for Policy Studies estimated that a fire resulting from an accident (such as the dropping of a cask of fuel that cracks the floor, draining the pool) or from sabotage could spread radioactivity over an area of 1,200 to 16,000 square miles. Even the low end of that range would be much larger than the area affected by the 1986 Chernobyl accident. The National Research Council projected in 1997 that such an incident could result in 54,000 to 143,000 extra cancer deaths, 0.5 to 1.5 million acres of agricultural land condemned, and evacuation costs of $117 to $566 billion.

High-level radioactive wastes in Europe are commonly vitrified before storage. The process, which produces safer-to-handle glassy hunks of still highly radioactive material, is finding support in the United States. But vitrified wastes, like all reactor wastes, generate lots of heat and have to be kept in an air-cooled containment area for as long as forty to fifty years before they have lost enough heat that they can be buried safely deep in the earth. It's another of those many ironies of climate control: at one end of the electrical grid, an air conditioner extracts heat from a home; at the other, radioactive elements that once kept that air conditioner running now have to be air- or water-cooled themselves, perhaps long after that air-conditioned house is gone. Waste ^{137}C has a half-life of thirty years; that is, within any thirty-year period, half of it decays, and radioactivity reduces by half. Other radioactive elements hang around in spent fuel much longer. The most problematic is plutonium, with stable isotopes ranging in half-life from eighty-eight years to eighty-eight million years. The isotope of greatest concern is ^{239}Pu, with a 24,000-year half-life. The national target is safe storage for 210,000 years. Those two thousand centuries are longer than Homo sapiens has been around. In just one-twentieth as many years, humans went from the hunter-gatherer life to the nuclear age.

Vexing problems continue to multiply in the world of nuclear energy, yet for every problem, proponents offer a seemingly miraculous solution. One of the most seductive approaches to dealing with radioactive wastes is reprocessing, which is designed to increase fuel supplies and attack the waste problem simultaneously. By chemically teasing apart the elements in spent fuel, it is possible to recover uranium and plutonium that can be used as fuel in conventional or breeder reactors, while leaving wastes that are easier to handle than was the original spent fuel. Or that's how it's supposed to work. France has long had reprocessing facilities capable of handling 1,600 tons of spent fuel annually. But more than 30,000 tons of potentially usable reprocessed uranium fuel have piled up in storage. Meanwhile, France has accumulated an estimated 630,000 cubic meters of radioactive wastes sitting at two sites, leaking radioactivity, with little prospect of finding a new home anytime soon.

In producing huge inventories of plutonium and other radioactive wastes, we have taken on a (literally) heavy responsibility, with no plan for how to deal with it in either the near or the distant future. Now respected people are insisting that we enter a new crash program to expand nuclear power, multiplying the burden of such wastes manyfold. And if we build twenty-first-century society around nuclear power, there's no backing out. We will be committing the human species, probably for the remainder of its run here on Earth, to the care and feeding of a complex, high-maintenance infrastructure that cannot fail if we are to keep the ecosphere safe from radioactive wastes. Durable physical barriers are only one part of the problem. Even if they work, how can a "Keep Out!" warning be communicated down through the millennia, into the mists of a distant future when, if our species is still around, it is likely no one will use any language, script, or medium known today? And how to ensure that whatever message we use will repel and not attract the curious? Scientists, artists, linguists, and various other breeds of thinkers have been pondering that problem for years, coming up with little beyond ideas like using bright colors and jagged graphics.

Meanwhile, back here in this lifetime, natural forces will continue to offer reminders of how fragile our own efforts at climate control are. During a 2006 European heat wave, the government of France, a nation that gets 80 percent of its electricity from nuclear plants, announced that "to guarantee the provision of electricity for the country," reactors would be allowed to discharge waste water at higher temperatures than are normally allowed under environmental laws. That alarmed environmentalists, who pointed to a government finding that when heated water was drained from the reactors three years earlier, "hot water temperatures might have led to high concentrations of ammoniac [ammonium chloride], which is potentially toxic for the rivers' fauna." Antinuclear activist Stephane Lhomme told reporters, "Global warming is showing the limits of nuclear power plants."

On the last day of July 2006, American Electric Power Co. shut down one of two nuclear reactors in Bridgman, Michigan, that supply electricity to Chicago and other areas. Water from Lake Michigan normally used to cool the reactor cores had itself become overheated by the summer's blazing temperatures, and inside the containment building, it was 120°. The following summer, the Tennessee Valley Authority shut down the Browns Ferry reactor in northern Alabama when its water supply from the Tennessee River reached 90°, an ineffective temperature for cooling the core. A utility spokesperson said, "It's the hottest in 20 years. We don't believe we've ever shut down a nuclear unit because of river temperature." Both incidents were especially untimely, occurring during periods of peak power usage by air-conditioning; record temperatures first pushed air-conditioning demand and electricity usage to the limit and then shut them off. Future warming of the atmosphere by greenhouse gases is expected also to raise the temperatures of rivers and other water bodies that supply cooling water for nuclear energy, so the frequency of plant shutdowns during times of peak air-conditioning demand could rise in coming years.

Can non-fossil fuels do a better job of running vehicle air-conditioning? That can happen only if alternative sources are up to the job of running the vehicles themselves. I estimated in chapter 2 that an all-electric vehicle fleet as large as our current fleet would compete strongly with air-conditioning of buildings for available power, and prospects are poor for supplying both from renewable electricity sources. The other well-hyped alternative involves fuels produced from grains and other biomass. In 2008, America produced 9 billion gallons of ethanol fuel, enough to displace only 4 percent of gasoline consumption—not enough to run the air-conditioning in the nation's cars and trucks, let alone propel them. Biodiesel production is negligible compared to consumption of the crude-oil-derived diesel fuel that it is meant to replace.

Currently, the chief source of biofuel in this country is grains, primarily corn. That's because corn produces more chemical energy per acre than do other temperate-zone grain crops. Suppose the entire land area now used to grow all crops in the United States—350 million acres—were sown to nothing but corn, and that the crop produced the current national yield of 155 bushels per acre (an unrealistically optimistic assumption, because much of the country's land and climate is not good for growing corn). That would provide enough fuel ethanol to substitute for about 70 percent of current gasoline consumption and require vast inputs of fossil fuels for cultivating, planting, fertilizing, spraying, and harvesting the corn—leaving no cropland for growing food. Grain-based ethanol is by no means "green," its chief ecological impact being the destruction of irreplaceable soil that occurs wherever grains are grown. Congress has dictated that grain ethanol production is to rise to 15 billion gallons by 2015. That would probably cover mobile air-conditioning demand that year, but it would be left to fossil fuels and other still-on-the-drawing-board biofuels to supply the power needed to move cars and trucks down the road.

A U.S. Department of Energy/Department of Agriculture blueprint for supplanting only 30 percent of the country's petroleum consumption with biofuels by 2030 depends heavily on emerging technology. The program, if pursued, would not only consume vast quantities of intensively cropped grain but also would strip 75 percent of crop residues like corn stalks and wheat straw from the soil after harvest. That will further deplete the organic-matter content of farm soils. The plan would also require intensive use of forest lands, also depleting their organic matter. And the DOE/USDA projections depend on the extraordinarily optimistic assumptions that grain crop yields can be increased as much as 50 percent, that perennial biomass crops will expand rapidly, that forest productivity will go up, and that energetically and economically efficient methods can be developed to convert cellulose to ethanol on very large scales. None of those bets is even close to being a sure thing.

9

COMING OUT OF THE COLD

The basic tenet of the adaptive model is that building occupants are not simply passive *recipients of their building's internal thermal environment . . . but rather, they play an* active *role in creating their own thermal preferences.*

—Richard de Dear and Gail Brager, 2001

To survey the full range of techniques that have been proposed for more ecologically sound thermal comfort could, and indeed does, occupy entire books. Any brief search for low-energy cooling will turn up a bewildering number and variety of devices and ideas. Some seem promising, some make slightly exaggerated claims, and some are simply blowing hot air, helping to inflate what entrepreneur Eric Janszen predicts will be a green-energy bubble to rival in size the dot-com and housing bubbles.

In this final chapter, I will briefly describe some of the more widely discussed or implemented proposals. They fall into three general categories: reducing the burden on air-conditioning systems; incorporating conventional, compressor-driven air-conditioning into less energy-intensive comfort systems; and ditching air-conditioning entirely. Many of these approaches are simply more energy-efficient methods of cooling and/or dehumidification; as such, they are vulnerable to rebound and backfire and can never be more than partial solutions. Incorporating technical efficiency and energy conservation into indoor climate systems cannot by itself ensure progress toward sustainability in the society at large. We will need comprehensive limits on consumption, in the form of material and energy limits on individuals, households, and businesses, with no licenses to pollute being sold. The limits will have to be agreed upon and enforced society-wide. Most of the tactics that follow represent steps toward reversing wasteful energy use, but each then will have to be incorporated into a strategy for pressing on much further. In the process, we may, to our surprise, rediscover some of the enjoyment of life that has been lost in the age of air-conditioning.

LIGHTENING THE LOAD

Most current approaches to less energy-intensive thermal comfort do take some form of refrigerated air as an assumption. In attempting to shrink air-conditioning's environmental footprint, researchers have proposed making all kinds of adjustments, not just to thermostats but to homes, offices, and even human beings.

Size and speed

Specialists say that one simple way to reduce energy use for cooling is to halt the practice of oversizing central air-conditioning systems. Some oversizing is necessary to cover general uncertainty; if a system is sized conservatively and then the customer complains of poor cooling, replacement is costly. But contractors often over-oversize just to be doubly sure, and because excess power can hide the faults of a system that is inherently inefficient or poorly maintained. A wrong-sized compressor uses more energy than a properly sized one doing the same job, so total consumption could be reduced by cutting back on the degree of padding built into systems. One way to save energy while reducing the risk of a poor fit between the equipment and the job is to use a system that can behave as if it's either large or small, depending on cooling demand. Analyses have shown energy savings of up to 25 percent if cooling power can be varied. Variable-speed electric motors have been available in America for a century, but are not yet common in air-conditioning; in contrast, Japanese manufacturers have been using them for two decades.

People cooling

A major factor in making the United States the world's top user of cooling energy was the big conversion, starting in the 1960s, from window units to central air-conditioning systems. The very different "people cooling" approach prevalent in most other nations is inefficient when viewed through the traditional American lens of comprehensive climate control, and when done haphazardly, it can even be less efficient than some of the more advanced central air systems discussed here; however, targeting people rather than buildings for cooling can consume much less energy when done properly.

Programmable thermostats have long been advocated as a way of easing off on cooling power during daytime hours when a house's occupants are all away at work or at night when employees of an office or a store are at home. However, the benefits can be canceled out by too-aggressive cooling at other times. Some central air systems in larger buildings and even in some large homes can be adjusted room by room or area by area to provide cooling only where needed and save energy. But for people cooling as an alternative to house cooling, the old-fashioned window air conditioner remains the most readily available option. Surveys of apartment residents in California and New Jersey, all of whom had window air conditioners, found that those who used manual controls consumed much less electricity for cooling than did those who kept their units on an automatic setting. The energy savings were realized even when fan and compressor controls were used in ways that were based on a misunderstanding of the device's workings. For example, many people used the thermostat as if it were a "cold valve" that produced air of varying temperature, when it actually turns the flow of uniformly cooled air on and off; nevertheless, those who did so realized some of the greater energy savings in the study. Other studies have shown that electrical consumption can be reduced considerably by installing indicator lights that show when temperatures have dropped enough outdoors to suggest turning off the cooling and opening windows. It is technically feasible to accomplish the same thing automatically.

In America, window units long ago acquired an association with lower socioeconomic status, a stigma that may be hard to overcome even in difficult economic times. Furthermore, closing off and cooling only occupied spaces just before and during the time they are occupied would require a significant change of attitude toward comfort. To idle vast inventories of expensive central-air equipment only to purchase new room units and designate intermittent "cooling centers" within our own homes would strike many Americans as absurd. But adjusting, modifying, or replacing central-air systems to support intermittent and/or spot cooling would save many billions of kilowatt-hours with little actual reduction in quality of life. In most of the rest of the world, attitudes are more in line with those of Japan, where, according to analysts Haruyuki Fujii and Loren Lutzenhiser, an emphasis on "people cooling" means that "ducted air-conditioning systems are generally not desired." In most countries, the most common room air-conditioning technology is the "split" unit, which has its hot components outside the house and its cool-air outlets inside, with only small refrigerant pipes running through the wall. It is more efficient than a window unit.

Where continuous air-conditioning of a whole building is currently being done, we need to devise and encourage modifications and operating strategies that localize the operation of refrigeration equipment in both space and time. Note, however, that a large, exposed thermal mass (like a concrete floor) can store and release heat, thwarting attempts to cool intermittently. And spot cooling is more easily prescribed than done, because air and heat don't like to stay in one spot. But insulating ceilings, cork or carpet on the floor, lightweight or cloth-covered furniture, and insulating partitions can help allow a small space to respond quickly to intermittent cooling.

The contemporary counterpart to our ancestors' caves—the basement—can often provide cool refuge in summer. The surrounding earth has a huge capacity to absorb heat from intruding warm air. If a basement is not damp but does suffer from ambient humidity that causes discomfort when it's warm in summer, an air conditioner set just low enough to cool the air even without mechanical dehumidification can make occupants comfortable for a low energy investment. That works because it takes less energy to remove sensible heat than to remove latent heat. And except under extreme temperatures and humidities, people are more sensitive to temperature variation than they are to humidity variation. We may say, "It's not the heat! It's the humidity!" but given the same amount of energy expended, a temperature improvement usually increases our sense of comfort more than does a reduction in humidity.

Reducing the indoor heat load

In most cases, it is sunlight and hot, often humid outside air that puts the biggest load on an air-conditioning system. Indoors, human bodies also generate heat and exhale warm, humid air. In large numbers, as in a gym during a high school basketball game, people can become the dominant heat load. But that's not the case on a quiet summer evening at home. The person in the kitchen chopping vegetables is imposing a heat load equivalent to about 200 watts, about half of it sensible heat, the other half associated with increased humidity from exhalation and perspiration. The couch potato in the other room is putting out only about 100 watts (see Table 1). Pound for pound, the family dog generates more heat than the human occupants because of its higher metabolic rate. But as the sun sets and incandescent lightbulbs are switched on, lighting's waste heat can easily exceed the total heat being generated biologically. The heat load from lightbulbs alone can be comparable to that imposed by a houseful of guests. Changing to compact fluorescent bulbs, which, compared with incandescent bulbs, generate only 30 percent as much heat for a given amount of illumination, could save energy not only for lighting but for cooling as well. Houses or other buildings with extensive natural light will save even more, if windows are tight and the light is not coming in the form of direct solar rays.

A substantial portion of the heat that burdens an air-conditioning system comes from our other household possessions, and many double reductions—in wattage used to run devices and again in wattage no longer needed to remove their output of heat—could be realized. Inefficiency in a refrigerator or freezer is passed on to the air conditioner to remove from the house. Dishwashers, showers, stoves, coffee makers, and clothes washers and dryers that do not vent outdoors add both sensible and latent heat. Washers that do a better job of spinning moisture out of clothing can substantially reduce the heat and humidity produced by dryers. Most clothes dryers expel much of their heat to the outdoors, but no indoor heat at all is generated when solar clothesline "technology" is employed. Computers, televisions, and other electronic devices suck energy from the wall socket and expel it as heat. Any reduction in summertime home energy use brings with it reduced demand for cooling. It is rarely if ever necessary to run a home air-conditioning system to protect the household's computers. Laptop models, for example, are built to function well with ambient temperatures of 100° or higher.

The superheated air that an air-conditioning unit normally releases into a house's backyard can, conceivably, be captured and used. The best medium for storing the heat is good old water; therefore, the waste heat is most often used to couple air-conditioning with a water-heating system. Typical savings on an annual water-heating bill in the southeastern United States are estimated at $50 to $150.

Fans

Sustained air movement within a cooled space doesn't lower the temperature, but it does enhance the human body's ability to shed heat. ASHRAE standards decree that an indoor "breeze" of 1.8 miles per hour improves comfort enough to allow the thermostat to be set 4.7° higher. In industrial, warehouse, and living spaces (or, someday, "paperless offices"?) not subject to the flying-paper problem, air speed can be cranked up to well above two miles per hour, and thermostats can be turned up several degrees. Fans can be used in either naturally ventilated or air-conditioned spaces. But if used to supplement air-conditioning, ceiling fans can in some situations cancel out the energy savings by bringing hot air down into the occupied space. Recall the air conditioner factory described in chapter 5; there, efficient cooling depended on keeping the air still.

Cool roofs and vegetation

Urban heat-island effects have burdened air-conditioning systems with increasingly heavier loads as the years go by, and much research has gone into reversing that trend. Hashem Akbari of the Lawrence Berkeley National Laboratory and his colleagues have long advocated increasing the reflectivity of whole cities, so that the sun's energy is sent back to the sky and not absorbed. They point out that energy can be saved both directly, by keeping heat out of individual buildings, and indirectly, by keeping the entire city cooler and thus putting less stress on cooling systems. Lighter-colored, more reflective roofs and exterior walls have demonstrated summertime energy savings of 15 to 40 percent in individual buildings. But those savings are realized mostly in warmer climates, because replacing a dark, heat-absorbing roof with a more reflective roof means giving up a source of solar warming in winter. In a 2001 review of cool roofs, Akbari skirted a direct comparison between summer energy savings and winter losses by using different units of measurement. After citing an impressive projected savings of 10 billion kilowatt-hours per year for cooling nationwide if all residences and commercial buildings had cool roofs, he estimated the resulting increase in natural gas consumption for heating at 26 trillion Btu annually. Translated into common energy units, the net savings are only a little over 2 billion kilowatt-hours, equivalent to less than one-half of 1 percent of air-conditioning energy consumption that year. Clearly, cool roofs will have their greatest positive impact in warmer climates.

Before World War II, U.S. urban areas were often cooler than the surrounding countryside because of extensive tree planting in cities and the spread of agriculture in formerly forested lands in the East and Midwest. When air-conditioning replaced shade as a primary means of cooling, cities became universally hotter than rural areas. Tree planting is now widely advocated as an anti-heat-island measure. Recorded reductions in cooling energy through direct shading of buildings have reached as high as 50 percent in very hot climates. Trees can reduce heating energy as well, by shielding buildings from winter winds. Furthermore, planting and maintaining trees in densely populated urban areas contributes in a way to social justice: in addition to improving a city's aesthetic appeal, trees also provide shade and evaporative cooling of the immediately surrounding air, giving some relief to people nearby who cannot afford air-conditioning. Through those same mechanisms, trees have a general indirect cooling effect on the city as a whole. Overall temperature reductions by trees have been estimated at 1.8° to 5.4° for ten U.S. cities. Trees also reduce air pollution both directly and through cooling.

For maximum conservation, it is recommended that trees be positioned to shade west, southwest, and south windows, so that at maturity, the edge of the tree's canopy comes close to the house. Deciduous trees, which do not shade the house in winter, are also beneficial when situated on the south side. Along the south side of our own house in Kansas, there is not enough space for a shade tree, so we grow a strip of the giant reed *Arundo donax*. This perennial, thick-stemmed, fast-growing grass with cornlike leaves emerges in spring and shades most of the house's south face during the hot period of July, August, and September. It covers even the second-story windows and dies back to ground level with the first frost, allowing winter sunlight to warm the side of the house.

Covering a roof with vegetation shields it from the more intense solar radiation to which it is subject, dampens temperature fluctuations through the thermal mass of the soil, and cools further by transpiring water, which absorbs heat and escapes as vapor. Energy savings by "green roofs" vary widely, depending on local climatic conditions and the method and type of vegetation used. One study found that ivy had a stronger cooling effect than grass planting, because ivy had a larger leaf area for transpiring water. Topping a house or office with a layer of soil and growing plants, however, is no simple matter. Green roofs must be designed and maintained with great care to avoid the host of easily imagined problems that can plague them. They are expensive and can consume a lot of water, yet can dramatically reduce energy consumption under the right conditions. And unlike reflective roofs, they don't reject solar warmth in winter.

Signals and incentives

Utility companies have tried a range of strategies to reduce summer power demand. In one such strategy, called "direct load control," the electric company sends electric signals that automatically prompt customers' thermostats to turn their air conditioners off or on during peak-demand periods. The off-and-on cycle is short—a fraction of an hour—and the unit's fan continues to run, so that customers are not alerted when cooling is or is not happening. Statistics from a New Jersey utility's direct load control program showed that it was effective in reducing peak loads while causing little discomfort among customers. But direct load programs are not even intended to reduce energy consumption, argues Ray Dean: "This utility ploy may reduce power demand (in the short run only), but it's very hard to believe it saves electrical energy. What it really does is to force customers to help increase utility profits. It allows utilities to shift some of their fossil fuel consumption from expensive (and cleaner) natural gas to cheap (and dirtier) coal. It doesn't save electrical energy because brief curtailments of local air conditioner operation are followed immediately by equal and opposite extra operation of that same cooling equipment." Dean doubts that such schemes can even have much effect on peak demand: "As time passes, people will adapt to such policies, and whenever a new house is built or an air-conditioning retrofit is introduced, that awareness will motivate installation of equipment with larger capacity."

In effect, says Dean, utilities are trying to shift the burden of excess capacity from themselves to the customer. Yolande Strengers also opposes direct load control on the grounds that it is "paternalistic" and assumes "that either failing to meet demand, and/or asking consumers to change their practices that create that demand, is unacceptable." She compared a direct control strategy implemented in Australia to an alternative voluntary program that dramatically raises a customer's electric rates during peak-demand hours in exchange for reducing rates at other times (and an advance rebate of one hundred Australian dollars). The program has driven peak demand down. A utility employee told Strengers, "Before we started the pricing trial, I thought customers would be prepared to pay AU$1.67 per kilowatt-hour," compared with the usual price of around thirteen cents, "to use their air conditioner because of the comfort factor. But the evidence suggests that the majority of customers are willing to make that sacrifice" to save on their electricity bills. The rate-raising approach, argues Strengers, "encourage[s] a more cooperative approach, engaging customers with their demand in order to ensure constant supply."

Economists would like to believe that as fossil fuel reserves dwindle and energy costs rise in the future, people will voluntarily reduce energy consumption. But a gradual increase in overall rates is unlikely to have much effect. In the United States, the average household spends $230 per year on electricity for air-conditioning. Double the rates, and people would still be spending less than one penny from each dollar of a median income to run their air-conditioning as much as ever. Where people spend more—as in Florida, where more than $600 per household per year already goes to power air-conditioning—steep rate increases might have an effect. They would certainly hurt households living in poverty.

Rather than general rate increases or peak-hour increases like that in Australia, a "progressive" price incentive, analogous to the progressive income tax, is more effective in reducing total energy consumption, not just peak demand. In such a scheme, escalating rates are imposed per kilowatt-hour based on total monthly electricity use. As the energy consumed by a household surpasses each of a series of thresholds, the cost per kilowatt-hour increases. Used by utilities in places as diverse as California and India, progressive pricing is a strong deterrent to consumption. Dean has called for such policies in his state but says, "Whenever I suggest this policy to a Kansas utility person, they know what I'm talking about, but they try to disparage the practice with statements like, 'It'll never work because customers don't care about saving money.' But people do care. The problem for the utility is that this policy might actually reduce the consumption of energy," which would cut into their profits. It would also draw protests from high-consuming customers.

Peak-rate and direct-load programs run by electric utilities are cream-skimming operations, relieving the companies of their biggest headache—excessive peak power demand—while not aiming directly at lower total energy consumption. Strengers noted that more energy-efficient home construction that would allow natural ventilation at off-peak hours would, from the point of view of the utility officials she interviewed, turn good customers into deadbeats: "Although frustrated, several interviewees admitted that improving the efficiency of the built environment would be 'counterproductive' for electric utilities, because, as one interviewee stated, 'it results in lower consumption at times when the network is not constrained,' which negatively impacts on the profit margin of the business." No business operator wants to invest in extra production capacity that's used only a small percentage of the time, and all businesses want to sell more of their products. In a full-blown hot-weather crisis, when peak demand surpasses its capacity, a company has to buy power on the spot market, in which they negotiate with another utility for quick delivery of energy, usually at a very high cost. Utilities are now encouraging summer peak-load reductions for the same reason they once promoted and sold air conditioners: to equalize demand across the seasons. If that steady "base" demand rises over time, private utility owners are happy to build new power plants. They just want the plants they build to be as fully utilized as possible. Utility-sponsored peak-reduction campaigns or rate incentives, while improving profitability, may even worsen the climate crisis. If customers shift their electricity use from afternoon or early evening to off-peak hours, they could even increase total greenhouse-gas emissions. The reason: at peak hours, the additional demand is met primarily by natural gas–fired plants, whereas the majority of base-load generation that fills off-peak demand comes from coal plants, which pump out twice as much carbon dioxide per kilowatt-hour and more of other pollutants than do gas-powered plants. To achieve significant energy conservation and greenhouse-gas reductions, public policies or campaigns will have to target total electrical *energy* (kilowatt-hour) consumption, not just peak-*power* (kilowatt) reduction.

Cooling centers

Increasingly frequent heat waves have led to calls for universal access to home air-conditioning as a public health measure. *Heat Wave* author Eric Klinenberg testified before the California state senate that "ideally, we would democratize access to home air-conditioning so that everyone had what most people take for granted." Klinenberg speaks for many who see heat becoming a social-justice issue, and see air-conditioning as a defense against hot periods that appear to be growing more frequent and hotter every year. Sitting in a comfortable home or office, it's easy to criticize those calling for expanded use of home air conditioners since more air-conditioning will neither relieve the economic stresses that make people vulnerable to heat nor address the underlying climatic causes of heat waves. But until all households have safe, decent living conditions, there is a desperate need for an alternative to the current system, which rations heat relief based on ability to pay. Here is where air-conditioning, a technology that has done as much as any to increase social isolation, could help bring communities together in a future age of economic and climatic stress. The public health concept of emergency cooling centers could be combined with something like the tried-and-true strategy of drawing people into chilled movie theaters and retail stores. The result might be attractive, spacious, moderately climate-controlled public spaces—"cool parks"—where people could congregate for shade, thermal relief, and social, cultural, and political interaction. If the communal cooled space were substituted for a sufficient amount of privately cooled space, it would be a net energy saver because the same number of people could be kept comfortable in a smaller space.

At home and on the road

All efforts to cool living and commercial spaces without refrigeration will have much greater chances of success if housing and retail square footage not only stops growing but begins shrinking to more reasonable proportions. A few local governments in places that include Atlanta, Austin, Pitkin County, Colorado (which contains Aspen), and even Los Angeles have put various types of house-size ordinances in place, limiting either absolute square footage or floor space as a percentage of lot space. But those ordinances still leave plenty of room for structures significantly larger than today's average new house.

If the United States accepts the challenge of deeply reducing resource use and greenhouse emissions, one of the many actions to be taken would be to direct job growth toward providing the necessities of life and away from buying and selling goods that no one really needs or even wants. The breakneck growth of the retail sector that ended in 2008–09 need not resume; the country can survive without it. Reducing superfluous consumption would be a painless way to reduce retail's many environmental impacts, including those caused by air-conditioning the country's vast retail square footage.

Strategies to reduce the energy used for cooling in transportation cannot be separated from the need for deep cuts in our dependence on private vehicles. Until public and human-powered transportation can become a serious substitute for private cars and until we can reduce the number of cars on the road and the miles they travel, efforts to reduce the energy used by mobile air conditioners will go for naught. Technological improvements will only gnaw at the margins. For example, the National Renewable Energy Laboratory estimates that cooling vehicle occupants directly by cooling their seats could save just 7.5 percent of the energy normally used by mobile air-conditioning. The EPA, with an international consortium, established the Improved Mobile Air-Conditioning Project in 1998, and by 2004, the group had come up with a list of recommended changes that were projected to cut mobile air-conditioning's fuel consumption by 30 percent and its refrigerant emissions by half. Recommendations included improved refrigeration equipment, tightened leaks, solar-reflective paints and windows, and improved ventilation. This array of improvements has not yet been made, and with automakers' economic distress, prospects for such improvements may have dimmed further. But the state of California has mandated those reflective windows (as discussed in chapter 4). Meanwhile, much gas and money can be saved by keeping in mind that every mile not driven and every minute a car is not idled saves much more fuel per mile or minute than does any equipment improvement.

AIR-CONDITIONING EVOLVES

Improvements in the operating efficiency of air-conditioning equipment could lead, in theory, to reduced resource use; however, as we have seen, those gains won't be realized without more sweeping changes. Elizabeth Shove laid out the situation well: "The promotion of technical efficiency is by far the most common policy response, yet it is one that internalizes and takes for granted those features of indoor climate change that are most problematic . . . technological changes do not simply meet demand, they also help construct and sustain it." Technologies designed to transform air-conditioning can improve the odds of significant energy savings, but technology alone can't bring environmental impact down to a sustainable level.

Heat wheels

To avoid sick-building problems, designers must provide adequate ventilation from outdoors, and that increases sensible and latent heat loads. A heat exchanger helps solve that problem by transferring heat from incoming ventilation air to outgoing exhaust air. In places where hot outdoor air is relatively dry, the best choice is a simple sensible-heat-only type of heat exchanger. Sensible-heat exchangers reduce temperature, not humidity; therefore, they need transfer only energy, not matter. Supply and exhaust air streams can be completely isolated from each other. Complete physical isolation is important in buildings requiring high levels of ventilation. In hospitals and chemical-processing facilities exhaust pollutants must not leak back into incoming ventilation. Where there is no health risk, a "heat wheel"—a thick, porous disk through which air can move—is useful. A sensible-heat wheel uses temperature difference to move sensible heat from a hot-air passage to a cold-air passage. Enthalpy, or "desiccant," heat wheels move latent heat from a humid air passage to a dry air passage; they are coated with solid desiccants such as silica gel. In the simplest method, hot, humid incoming air passes through the slow-turning wheel, and cooler, drier exhaust air passes through the other side of the wheel to remove the heat or moisture that was extracted from the incoming air.

Solar devices to dry out the desiccants in wheels have been developed, and they can achieve significant energy savings.

Solar air-conditioning

Using solar radiation to cool buildings is a true jujitsu move, because it is at precisely those times when summer sunlight is strongest that it can be used to satisfy peak demand for cooling. Electricity generated by photovoltaic panels can power traditional air-conditioning, or the sun can be used as a heat source to run what are called absorption and adsorption (together known as "sorption") systems. Sorption cooling has been understood for a century, and was used in refrigeration before more efficient electrically driven compressors were developed. Even Albert Einstein, with colleague Leo Szilárd, harnessed his intellect to the problem of absorption refrigeration, applying for a patent in 1933. Detailed descriptions of such systems are complex, but their great appeal lies in the absence of moving parts or any ozone-depleting or planet-warming refrigerants. Any sorption cycle involves the heating of a liquid solution or a solid material to boil off a vapor, so energy, as always, must be invested. Sorption systems are especially well suited to running on solar energy as the heat source, however.

Since sorption refrigeration is driven by heat flow rather than by mechanical power, it cannot raise temperatures to the high levels achieved by mechanical refrigeration; therefore, the laws of thermodynamics dictate that it will be less efficient. Because of that, sorption air conditioners require a lot of heat input in order to move heat out of a space, and they are large and costly. But because they run on inexhaustible sunlight, solar-based sorption systems use fossil energy resources only in their production, not in their operation. As an alternative to a solar-powered sorption machine, a photovoltaic-powered conventional air conditioner might actually be more energy-efficient, but it would have the refrigerant problem. Either type of system would be a huge improvement over fossil-powered central air.

The largest facility in America currently being cooled by a solar-powered absorption system is a nondescript, multistory concrete building near the corner of Seventh Avenue and Camelback Road in

Phoenix. Cooling of the building, which houses a self-storage business, starts with heating of water to 220° in solar collectors that cover five thousand square feet on the roof. A six-thousand-gallon insulated tank stores the hot water, which provides energy to run a thirty-ton absorption unit housed in a shipping container in the parking lot. The 140° air temperature inside the container makes the 114° parking lot seem cool—and that makes its ability to provide cool air to the building seem all the more like magic. "The system was shipped from Austria just as you see it now, in that container," says Bob Livingston, president of Solar Energy Partners, Inc., in Phoenix, which provides sales and marketing for the system's developer, SOLID GmbH of Austria. "China has fifteen of these systems going now. We are way behind other countries." Government credits and grants make solar thermal systems much more attractive to businesses, he told me, but "the economy is going to have to get a lot better before this will be widely adopted." How expensive would electricity have to be before these systems could take off without subsidies? "It would have to go to twelve to fourteen cents a kilowatt-hour," Livingston estimates. That would a near-doubling of Arizona's current commercial/industrial rates, bringing them into line with rates in the New England states.

Different types of devices can be used effectively in complex combinations. Researchers in Florida have shown that a complex system employing a sensible-heat wheel, a moisture-extracting heat wheel, two evaporative coolers, and a hot-water coil connected to a hot-water solar collector can function as a solar-powered central air conditioner that uses 100 percent outside air throughout the cooling season. Ray Dean stresses that using all-outside-air is particularly important if you want to allow some people in the conditioned spaces to switch over to natural cooling by opening windows. Because all air is exhausted and never returns to the refrigeration equipment, there is no danger that humid air coming in through open windows will put a bigger burden on the air conditioner. If, when people open their windows, they also stop the cooling of their spaces (by closing a manual damper or setting their local thermostat to a high temperature), their space will automatically be completely removed from the central system's load. Such a system has a relatively high initial cost and a relatively high maintenance cost, but it has a low operating-energy cost. On especially humid days that are not extremely hot, indoor air can be made comfortable with very little mechanical cooling, and savings can be significant.

Ground-source heating and cooling

Ground-source heat pumps take advantage of the relatively constant, moderate temperatures and tremendous mass of the deep soil around and below a building. Using thermodynamically based mechanisms like those of an air conditioner, these systems transfer heat from the earth into a building or, in summer mode, move heat out of the building into the earth. In a closed-loop system, heat moves with an antifreeze solution that is pumped through horizontal or vertical pipes buried in the earth. In an open-loop system, groundwater or surface water is used to deliver or remove heat before being returned to the earth. Because cool soil or water can absorb waste heat more efficiently than can hot air, ground-source systems can slash the amount of energy used for air-conditioning. They have been popular for years in northern and central Europe, where their capacity for efficient heating is more important than their cooling capability.

Installation of ground-source heat pumps in the United States is growing by 12 percent per year, but the total energy delivered for heating and cooling still amounts to less than 2 percent of the energy consumed by conventional air-conditioning alone. Biodiversity Conservation India Limited has established a residential project called the BCIL Collective that is said to have the largest fully geothermal air-conditioning system in the world. One room in each residence (the master bedroom, naturally) is maintained at or below 75° with air cooled by circulation through "earth tunnels." No compressor-driven air-conditioning is used; the only operating energy used is for fans. At night, the system is aided by a mechanism that cools water by radiating heat to the night sky, then stores the water and circulates it through the geothermal conduits the next day ahead of the cooled air.

Evaporative cooling and wind towers

Hot, dry climates provide the widest array of cooling possibilities, because large quantities of heat can be removed from the air by evaporating water. Colonial India's "thermantidote," modern outdoor misting machines, green roofs, and even the traditional wet sheet over the doorway can take advantage of water's high heat of evaporation. Electric evaporative coolers that pull air through water-saturated pads remain popular in parts of the southwestern United States and are still much more common than air conditioners in India and many other arid or semiarid regions. Their fans use much less energy than is consumed by fans plus compressors in an air conditioner. However, evaporative coolers can also use up a lot of water in the dry climates where they work best; cooling an entire house for a day can convert hundreds of liters of precious water into vapor. Unfortunately, many locations where evaporative cooling is most effective are also plagued by water shortages.

The evaporative wind towers at Hyderabad's Green Business Centre are used to precool air before it runs through a conventional air-conditioning system. Where humidity is low—as in Hyderabad for as much as eight months of each year—moisture added to air in the towers cools it further and increases the downdraft. The wet surfaces cool the incoming air by about 20°, but also add humidity, which, if it reaches an oppressive level, must be dealt with by the air-conditioning system. Pressure changes created by the wind towers lower the temperature of the incoming air, so in some situations, they provide sufficient comfort even without using water or air-conditioning. Many towers, like the Persian and Arabian models of olden days, are rectangular, built of stone or concrete. A round tower made largely of fabric is being tested in Israel; it is capable of recirculating indoor air to achieve improved cooling. At the Torrent Research Centre in Ahmedabad, India, a dozen buildings are cooled by large arrays of wind towers. (Another much smaller portion of its square footage is air-conditioned.) The cooling system gets help from thermally massive concrete construction and reflective surfaces. During the hot, dry season, water is sprayed in the towers to enhance cooling; during the monsoon season, the water is turned off, but the towers, assisted by fans, continue to circulate air. With the lower temperatures of the monsoon season, air movement alone achieves adequate comfort levels. The system reportedly can bring summer temperatures of more than 110° down to 84°. With air-conditioned Indian office buildings using 280 to 500 kilowatt-hours per square meter annually and a national target now set at 140 kilowatt-hours for new air-conditioned buildings, the Torrent complex consumes only fifty-four kilowatt-hours.

Like wind towers, "solar chimneys" draw fresh outdoor air into a building, but in the reverse direction. A roof panel or other structure absorbs solar radiation and heats air in the chimney. The heated air rises out of the chimney, sucking hot indoor air out via induction. The cooling effect is achieved through that air movement. Simulations predict that a solar chimney can achieve three to six complete turnovers of the indoor air in a large room each hour. If the indoor air is hotter than outdoor air, or if air enters the room through evaporative pads or from a ground source (with or without air-conditioning), cooling can be achieved as well.

Building design and construction

As we have seen, better home insulation could save the United States 15 percent on heating and cooling energy combined. A return of some pre-air-conditioning construction features, such as large eaves and porches, would allow cooling systems to be switched off for much longer periods. Awnings, shades, shutters, screens, blinds, louvers, and porches (and porch swings) can be added to existing buildings. Greater thermal mass, especially through use of the earth below and surrounding the house for energy storage, can buffer temperature fluctuations. All such features tend to increase the quantities of materials and energy used in building or renovating a house, but their inclusion can save a much larger quantity of energy during the house's lifetime.

Cooling features that enhance ventilation are harder to integrate with air-conditioning systems because they increase the flow of warm and/or humid air into a building. Nevertheless, hybrid ventilation systems have been developed to combine natural with mechanical ventilation and/or either ventilation mode with limited refrigerated air-conditioning. Systems may be designed to switch between natural cooling and air-conditioning on a day-to-day basis or to vary both simultaneously. For example, in areas with large day-night temperature differences, an office building might be opened up and its air-conditioning turned off when evening arrives, to allow exhaust fan(s) to run through the night, drawing air into the building and cooling its thermal mass. Or fully natural ventilation might be used in all but the hottest, most humid, or windiest weather.

Obviously, natural-ventilation strategies work best in buildings designed for maximum cross-ventilation or with specific cooling features like wind towers or dehumidification equipment. Automatic control of vents, fans, and air-conditioning units is usually needed to make it all work, but thermally efficient windows that can be opened and closed easily by hand can be important. Operable windows permit considerably more outside air infiltration than do permanently sealed windows even when they are shut during hot or cold weather, so it may not be necessary or desirable to make all or even most windows in a building operable. With the right design, it doesn't take many open windows to create a substantial draft and beneficial cooling effect in upwind spaces; therefore, designers of large buildings could carefully select just a few strategically placed windows as the operable ones.

Objections to incorporating natural ventilation in hybrid systems center on uncomfortable drafts, exposure to air pollution and noise, and creation of openings in buildings that might invite burglary. In India, as in many countries, there are bars on almost every window in all but the poorest urban houses. Barred windows can present a somewhat grim face to the world, but if having them would mean that American homes and businesses could keep windows open for ventilation through more of the day, night, and year without fear of crime, bars might come to be seen as a green construction feature.

MAKE YOURSELF COMFORTABLE

Floods of money, knowledge, experience, and creativity are going into developing new technology for keeping cool, but the bigger the resulting temperature drop, in general, the more complex, expensive, and potentially fragile the technology. Although emerging methods generally use less energy than compressor-driven air-conditioning, no system for manipulating the indoor climate can avoid having an impact on the outdoor world. Furthermore, the proportional energy savings claimed by the myriad proposals for energy-saving methods of air-conditioning—15 percent here, 30 percent there—are not definitive numbers, and they do not combine in an additive way. The savings cannot simply be plugged into a simple formula for total savings as new efficiency and conservation practices are adopted. We should not make the mistake of the man who, in economist Herman Daly's telling, "bought a new stove that cut his fuel bill in half and then reasoned that he could cut his fuel bill to zero by buying another such stove!" And, of course, the complexity and expense of many of the systems described here means that they can be enjoyed only by a privileged minority of the world's people.

People and cooling systems must adapt to each other. Reasons for the popularity of refrigerated air-conditioning and the tendency of people to acquire it if they can afford it are obvious and widely recognized: the size and design of equipment needed to achieve a given comfort level in a given space are easily determined, and the equipment is easily accessible; technological improvements have made air-conditioning units highly effective, predictable, and reliable; simple adjustment of a thermostat provides very precise control of both temperature and humidity; and the cost of air-conditioning has become almost negligible relative to the cost of the building that it cools, while the market value of the house drops dramatically in most cases if it does not have air-conditioning. In contrast, natural ventilation or other nonrefrigerative means of achieving comfort are somewhat more unpredictable, require more human effort and attention, and accommodate greater fluctuations in temperature and humidity. Because such systems must be built to rely more on real-time energy fluxes and less on precisely metered-out fossil energy, and because they are not currently being mass-produced, their initial costs can be much higher. And some approaches consume water that may already be in tight supply.

Accepting, indeed enjoying, a broader definition of comfort can greatly broaden the range of options and reduce the cost to everyone. Any of the methods proposed or used for reducing air-conditioning's energy consumption can also be used independently of air-conditioning, if more flexible definitions of comfort are adopted. Architectural features, shades and awnings, trees, window and ceiling fans, whole-house exhaust fans, solar dehumidification, shade trees, reflective or planted roofs, ground-source cooling, large thermal mass, natural or cool lighting, community efforts to reduce the heat-island effect, and a focus on people cooling rather than space cooling all can be employed to improve summer comfort without refrigeration.

The buildings of the Torrent complex provide just one example of how to achieve non-air-conditioned comfort in a hot climate. But conditions are much harder to quantify in spaces ventilated with "natural" air, and that lack of mathematical precision, along with high initial costs, has slowed adoption. The architecture, construction, and real estate industries want replicable, consistent, concrete standards to aim for, and the chief source for such standards in this country is ASHRAE. The widely followed ASHRAE Standard 55 is designed to "produce thermal environmental conditions acceptable to 80% or more of the occupants within a space." Adaptive-comfort specialists Richard de Dear and Gail Brager argue, however, that "people who live or work in naturally ventilated buildings where they are able to open windows become used to thermal diversity that reflects local patterns of daily and seasonal climate variability" and tolerate much greater variation in warm indoor environments. Evidence has also accumulated to show that the temperature range that building occupants find comfortable is not fixed but varies along with outdoor temperatures in the present and recent past. It turns out that the temperatures preferred by most people, especially people who do not spend much time in air-conditioned environments, rise as outdoor temperatures rise. That has led to a broader "adaptive model" of thermal comfort, the idea that "building occupants are not simply *passive* recipients of their building's internal thermal environment . . . but rather, they play an *active* role in creating their own thermal preferences." Partly as a result of long struggles over the definition of thermal "acceptability," Standard

55 was expanded in 2004 to incorporate criteria for naturally ventilated buildings, including a broader and more flexible comfort zone.

Based on extensive data from "many thousands of human subjects
in building studies from around the world," de Dear and Brager recommended that under conditions of natural ventilation, and where people can adjust windows, fans, and their clothing, the indoor comfort
range be expanded to reach as high as 89° in hot weather. (ASHRAE
standards previously reached their upper limit at 80°, and that only under the lowest-humidity conditions.) Figure 4 shows the temperature
range predicted to be acceptable to at least 80 percent of people when

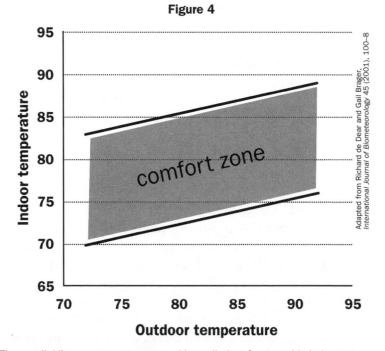

Figure 4

The parallel lines represent upper and lower limits of acceptable indoor temperature for 80 percent of subjects as the outdoor temperature rises from 72° to
92°. Limits are computed from equations published by Richard de Dear and Gail
Brager in 2001. Temperatures at which people feel comfortable depend on many
factors in addition to outdoor temperature: humidity, air movement, clothing,
activity level, and others. As outdoor temperature rises above the range shown,
individual comfort levels will continue to be influenced by those other factors.

outdoor temperatures are in a range between 70° and 95°. Note that in accord with the adaptive model of comfort, the range of acceptability rises as outdoor temperatures rise, at least up to the low nineties.

Studies have shown that with natural cooling, an air movement equivalent to a 2.2-mile-per-hour breeze can raise the comfortable temperature range by as much as 6°. It is also known that rapid fluctuation in air speed gives an added cooling sensation; although you may feel only a steady blowing, fans produce very rapid pulses, or tiny "gusts." In temperatures high enough to induce a little sweating, fans increase evaporation from the skin, providing yet more cooling.

Based on his analysis of data from tropical countries, Fergus Nicol of Oxford Brookes University in the United Kingdom has found an adaptive-comfort range of temperatures very similar to that published by de Dear and Brager. He has concluded that the relationship between temperature and comfort can never be fully described by mathematical or statistical models. Instead, he says, the relationship lies within a "black box"; therefore, any cooling technique must be judged by its practical outcome. He and others have emphasized that people adjust not only their clothing but also their activity level, posture, location, and state of mind in order to feel comfortable. Like de Dear and Brager, he stresses that occupants tolerate or even enjoy a much wider range of temperatures in a house or workplace where they themselves have a high degree of control over air movement and outside ventilation.

In a widely cited 2005 essay on the future of comfort, Heather Chappells and Elizabeth Shove made the point that "building occupants, engineers, and designers work with different languages of comfort. . . . The gulf between designers' detailed calculations on one hand and the effect of adding or removing a sweater remains wide." In the working world, they argue, allowing wider temperature fluctuations, along with greater flexibility in clothing and working hours, could allow much lower-energy cooling. Indeed, "variation itself is an important part of being comfortable and that far from representing a compromise, solutions based on these sorts of arguments are superior to those that deliver uniform conditions of 'thermal monotony.'"

What has been lost in the age of air-conditioning is as important as everything, good and bad, that has been gained. Among those things subtracted, thermal variation may be the least appreciated. Stephen Healy of the University of New South Wales believes that current comfort standards emphasize thermal monotony. While this monotony is "an innovative and inventive trajectory facilitated by science," argues Healy, it has replaced a much more valuable thermal variety. And he sees no clear path back to restoring variety: "With thermal monotony embedded in knowledge, institutions, people, buildings and machines, a successful strategy for promoting 'thermal adaptation' will have to be sophisticated and comprehensive." A positive feedback loop will help in that effort: after people have begun experiencing more thermal variety, science shows, they begin preferring perhaps even greater thermal variety.

In her 1979 book *Thermal Delight in Architecture,* Lisa Heschong argued that the thermal sense is important enough to be assigned status at the level of the other five senses, and that, like hearing, sight, taste, smell, and touch, the thermal sense requires stimulation by a variable background. Without the extremes, enjoyment of moderate conditions declines. After I have worked outdoors through a broiling-hot day, I find that walking into a supercooled office or grocery store is satisfying in the extreme—at first. Yet what I look forward to most is that moment at seven or nine or ten at night when, as I'm sitting on a porch or near a window, I feel that first slightly cool breeze come through. It can make all the preceding hours in the heat worthwhile. That, I realize, may make me seem a little daft, but the world provides a delicious spread of other thermal variations from which to choose—like that sunny summer afternoon when, in a flash, a storm closes in, and the atmosphere becomes one big evaporative cooler. Here, Heschong elaborates on the enjoyment of thermal extremes:

> [P]eople definitely seem to enjoy a range of temperatures. Indeed, they frequently seek out an extreme thermal environment for recreation or vacations. This must explain in part the love of Finns for their saunas and the Japanese for their scalding hot baths. Americans flock to beaches in the summer to bake in the sun and travel great distances in the winter to ski on frosty mountaintops. People relish the very hotness or coldness of

these places. We should note that all of these places have their opposites close at hand. The Finns make a practice of jumping from the sauna into a snowbank or cold lake. At the beach, after baking in the hot sun, there is the cold ocean to swim in. . . . The experience of one extreme is made more acute by contrast to the other.

DOING SOMETHING ABOUT THE WEATHER

It is noteworthy that even though much work is being done on more energy-efficient cooling technology in the United States, a disproportionately large part of that effort has been carried out in Europe and Japan, both in research and in practice. Summers are much milder across the greater part of those regions than they are in the U.S. Sun Belt, and that makes it much easier to achieve success with efficient technologies. In many climatically temperate, economically well-off countries, invention is necessity's mother; a host of high-tech, lower-energy devices and methods have been developed that can provide comfortable indoor conditions when the outdoor temperature is, say, a mildly muggy 88° or a dry 92°. Many such strategies would not stand up to truly harsh conditions; they can add comfort only when the weather is not exceedingly uncomfortable to begin with. In contrast, the only way to achieve significant reduction in resource use for cooling in places like central Arizona and south Florida is for those regions to adopt very tough restrictions that halt further growth and reduce the environmental footprint of the households and businesses already there. A reverse migration that reduces the populations of those areas may need to be the eventual goal. In fact, a turning point may already have been reached with the recent, sudden reversals of Florida's and Phoenix's population growth and the sudden suspension of southward migration across the country. Meanwhile, in those places across the globe where intense heat and humidity dominate, where people can't afford to pick up and move to cooler places, and where incomes cannot accommodate air-conditioning or other costly cooling methods, elegant, cheap, low-tech means of modifying some indoor conditions and adapting to others are employed daily.

We have built our shelter and transportation systems for the twentieth century, basing them mostly on eighteenth- and nineteenth-century concepts of an "empty world" into which we plan to grow indefinitely. The same ideas, technologies, and economic means with which we set in motion the looming ecological crisis cannot now be depended upon to keep the planet livable, indoors and out, in the twenty-first century.

The biggest obstacle to change is the way we have been constructing houses, offices, roads, cities, and suburbs for decades; all around us, the need for air-conditioning is literally set in concrete and steel. Extrication from the refrigerated world will mean extricating the economy and society from the entire assortment of predicaments in which they have become mired. Living in a less refrigerated country will mean actually living where we're located, not in a sterile, standardized environment. If that "somewhere" happens to be a desert or swamp, it will mean not living in the midst of a Disney Desert or hanging out in a Big Cypress Swamp Flea Market but living in a real desert or swamp and accepting the ecological responsibilities and pleasures that come with it. With the reduction in electricity demand, we can make things easier on everyone and bring the goal of a totally renewable energy supply closer to being within reach.

Utility companies can be made to focus on supplying necessary electricity as cleanly as possible and not on enriching themselves. Houses and vehicles can once again be viewed as means of shelter and transportation, not brazen investments; square footage and commuting time can shrink to a human scale. We can help one another get through heat waves rather than flee into cold isolation. With less climate control and more contact with the real ecosphere, we and our children might well feel healthier, physically and mentally. We'll be less estranged from neighbors and nature, and less likely to regard comfort as a mere commodity to be acquired in the marketplace. Comfort can be put back under the control of working people and not used by their employers as a tool to extract extra work. Regional differences can reflect variation in weather, ecosystems, and local cultures rather than empty political distinctions against a homogenized cultural background. We can leap off the consumption-efficiency-consumption treadmill and develop a healthier relationship with the technologies we do use, recognizing their technologically sophisticated, ecologically fragile nature.

By becoming more than anonymous x's and y's in a set of heat-load calculations, we can become more resilient human beings. And we'll need that resilience. The coming decades are going to be a harsh test of our ability to adapt and create, and we can't leave it to technology to bail us out this time.

NOTES

(All Internet addresses were accessed on October 3, 2009.)

Preface

ix *The End of the Long, Hot Summer:* Raymond Arsenault, "The End of the Long Hot Summer: The Air Conditioner and Southern Culture," *Journal of Southern History* 50 (1984), 597–628.

xi *We don't use the air conditioner:* Loren Lutzenhiser, "A Question of Control: Alternative Patterns of Room Air-Conditioner Use," *Energy and Buildings* 18 (1992), 193–200.

xi *Two books:* Gail Cooper, *Air-Conditioning America: Engineers and the Controlled Environment, 1900–1960* (Baltimore: Johns Hopkins University Press, 1998); Marsha Ackermann, *Cool Comfort: America's Romance with Air-Conditioning* (Washington, DC: Smithsonian Institution Press, 2002).

xii *Gwyn Prins's:* Gwyn Prins, "On Condis and Coolth," *Energy and Buildings* 18 (1992), 251–258.

xiii *new moralists:* Ackermann, *Cool Comfort,* 138–149.

Chapter 1

1 *Jim Roberts:* Jim Roberts, interview by author, Ft. Myers, FL, December 2008.

1 *Morris Udall:* "Central Arizona Project: Tapping Arizona's Last Water Hole," Congressman's Report, May 21, 1963.

1 *I thought I started hearing the heat:* Although Pueblo Grande borders Phoenix's airport, it was not airplanes that I was hearing. For a thorough description of Hohokam civilization in the area, see Todd Bostwick, *Beneath the Runways: Archaeology of Sky Harbor International Airport* (Phoenix: City of Phoenix, 2008).

3 *They proved it for more than a thousand years:* In the mid-1400s, after a millennium-long residence in this desert valley, the Hohokam suddenly

abandoned it. Hypotheses for the collapse of their civilization include drought, flood, war, or perhaps an influx from the north of immigrants fleeing their own droughts and economic collapse. According to Bostwick, the influx may not only have put pressure on resources; it may also have disrupted the elegant social organization, still only partly understood, that kept the elaborate canal system working. If the influx of population from the north was the straw that eventually broke the back of their sturdy civilization, there may be a lesson for modern Phoenix in the story of the Hohokam.

3 *air-conditioned capital of the world:* Arizona Republic, November 17, 1940, quoted by Michael Logan in *Desert Cities: The Environmental History of Phoenix and Tucson* (Pittsburgh: University of Pittsburgh Press, 2006), 145.

3 *Phoenecians do not move to new localities:* Ibid.

3 *raised temperatures 7.6°:* Jay Golden, "The Built Environment Induced Urban Heat-Island Effect in Rapidly Urbanizing Arid Regions—A Sustainable Urban Engineering Complexity," *Environmental Sciences* 1 (2004), 321–349. Throughout this book, I will express temperatures in degrees Fahrenheit unless otherwise noted.

3 *carbon dioxide dome:* Craig Idso, Sherwood Idso, and Robert Balling Jr., "An Intensive Two-Week Study of an Urban CO_2 Dome in Phoenix, Arizona, USA," *Atmospheric Environment* 35 (2001), 995–1000.

3 *In 2007, Phoenix endured a record-breaking twenty-eight days:* Fay Bowers, "More People, More Concrete, and Lots More Heat in Phoenix," *Christian Science Monitor*, August 30, 2007.

4 *on July 15–16, 2003:* Golden, "The Built Environment."

4 *six million by 2030:* projection by Maricopa County Planning and Development Department, http://www.maricopa.gov/planning/Resources/ Plans/docs/pdf/Rio_Verde_Foothills/pdf/06DEMO.pdf.

5 *Phoenix's air quality is well below:* Julie Cart, "Rapidly Growing Phoenix Finds Dust Unsettling," *Los Angeles Times*, September 7, 1999.

5 *The increase in childhood and adult obesity:* Golden, "The Built Environment."

5 *By 2008, Phoenix politicians:* Lisa Selin Davis, "Hope for a Desert Delinquent," *Grist*, May 13, 2008, http://www.grist.org/article/phoenix1.

7 *residents who are trying to live within tighter ecological limits:* East of central Phoenix, bordering the posh new developments of Scottsdale, is the Salt River Indian Reservation. Many members of the Salt River Pima-Maricopa Indian Community (http://www.srpmic-nsn.gov) live without air-conditioning, some voluntarily and others because they cannot afford

it. Most of the houses in the semiagricultural area have evaporative coolers, and some have window air conditioners. A few more upscale houses have central air. For insights into life in the community, see Janine Schipper, *Disappearing Desert* (Norman: University of Oklahoma Press, 2008).

7 *When Phoenix is done growing,:* Davis, "Hope for a Desert Delinquent."

8 *I practically grew up in Phoenix:* Quayle did grow up in Arizona. See http://en.wikiquote.org/wiki/Dan_Quayle.

8 *Critical to this effort was the assault:* Logan, *Desert Cities*, 145–46.

8 *On the one hand:* Ibid., 167–68.

9 *City officials declare that they have demand covered:* See the City of Phoenix's statement at http://www.phoenix.gov/WATER/drtmain.html.

9 *the city's tight water situation could undermine:* Golden, "The Built Environment."

10 *And xeriscaping:* Elizabeth Wentz and Patricia Gober, "Determinants of Small-Area Water Consumption for the City of Phoenix, Arizona," *Water Resources Management* 21 (2007), 1849–63.

10 *Disney Desert:* Larissa Larsen and Sharon Harlan, "Desert Dreamscapes: Residential Landscape Preference and Behavior," *Landscape and Urban Planning* 78 (2006), 85–100.

10 *We enjoy gardening:* quoted by Raymond Arsenault in "The End of the Long, Hot Summer."

10 *Seven years ago:* William Saletan, "Planet of the Indoor People," *Washington Post*, August 6, 2006.

11 *Projections show:* Susanna Eden and Sharon Megdal, "Water and Growth," in *Arizona's Rapid Growth and Development: Natural Resources and Infrastructure*, ed. Sharon Flanagan-Hyde et al. (Phoenix: 88th Arizona Town Hall, 2006).

11 *a decrease in water consumption:* Ibid.

11 *swimming pool:* Wentz and Gober, "Determinants of Small-Area Water Consumption."

12 *The National Climatic Data Center:* National Climatic Data Center, "Monthly State, Regional, and National Cooling Degree Days Weighted By Population," *Historical Climatology Series 5-2* (2008). See http://www .ncdc.noaa.gov/oa/documentlibrary/hcs/hcs.html.

12 *Heavy reliance on air-conditioning:* Throughout the city's older suburbs, it seems every roof has an evaporative cooler attached, and many are still used for a good part of the year. But residents will tell you that when relative humidity rises during the June 15–September 30 "monsoon" season, evaporative coolers can't supply the required level of comfort. Coolers

are rare in newer housing developments, and, as Dani Moore pointed out, they may even be banned.

12 *Consumption per person grew by 51 percent:* Data from U.S. Department of Energy, 2008, http://apps1.eere.energy.gov/states/state_information .cfm.

12 *takes 216 million gallons of fuel:* Data from National Renewable Energy Laboratory, "Air-Conditioning and Emissions," 2009, http://www.nrel .gov/vehiclesandfuels/ancillary_loads/ac_emissions.html.

12 *an average of 4,822 heating degree-days:* I calculated these figures by taking state data from National Climatic Data Center, "Monthly State, Regional, and National Cooling Degree Days," and 2007 state per capita energy consumption (from http://www.eia.doe.gov/cneaf/electricity/epa /epa_sprdshts.html), and computing the average, weighted by state populations as reported in the censuses of 1950 and 2000.

13 *about 3 percent of its residential electricity consumption is satisfied by solar:* U.S. Energy Information Administration, 2009, http://www.eia.doe.gov/ emeu/states/_seds.html.

13 *two hundred to four hundred megawatts:* Black & Veatch Corp., *Arizona Renewable Energy Assessment,* B&V Project Number 145888, September 2007.

13 *Before 1992, water flow in the Colorado River:* E.D. Andrews and L.A. Pizzi, "Origin of the Colorado River Experimental Flood in Grand Canyon," *Hydrological Sciences* 45 (2000), 607–27.

14 *there's nowhere else:* Alan Weisman, *The World Without Us* (New York: Thomas Dunne Books, 2007), 210.

15 *Phoenix has anywhere from a few hundred:* Michael Clancy and Casey Newton, "For First Time in Modern History, Phoenix May Be Losing People," *Arizona Republic,* January 12, 2009.

15 *eleven thousand residences:* Dianna Smith, "Ave Maria Project Sets Precedent for Conservation," *Naples Daily News,* December 19, 2004.

16 *Even as southwest Florida staggered:* Mitch Stacy, "Billionaire Makes College, Town Grow," Associated Press, May 9, 2009.

16 *the Everglades were still dying:* Michael Grunwald, *The Swamp: The Everglades, Florida, and the Politics of Paradise* (New York: Simon and Schuster, 2006), 358.

16 *Naples was a sleepy little town:* Collier Mosquito Control District, "Collier Mosquito Control District History," http://www.collier-mosquito .org/pdfs/historical_narrative.pdf.

17 *Naples had the second-highest per capita:* Bureau of Economic Analysis, 2009, http://www.bea.gov/regional/bearfacts/countybf.cfm.

17 *Although many Americans are poorer:* Mike Schneider, "Analysis: Resort Towns Attracted High-Earners," Associated Press, May 19, 2009.

17 *green and pink people:* John Rothchild, *Up for Grabs: A Trip Through Time and Space in the Sunshine State* (Gainesville: University Press of Florida, 1985), 93.

18 *car condos:* Dee-Ann Durbin, "Car Enthusiasts Turning to Condos," Associated Press, May 19, 2006.

18 *air-conditioned golf carts:* Otto Pohl, "In Hot Season, Sunbelt Golf Courses Stay Cool," *Wall Street Journal,* July 3, 2005.

18 *more than 150 golf courses:* Tim McDonald, "Golf Course Construction Slowdown? Not in Southwest Florida," Golf Publisher Syndications, February 28, 2007.

18 *Naples Big Cypress:* "Naples Big Cypress Market," company press release, November 2006, http://www.naplesbigcypress.com/NBC_press_release _november06.html.

19 Land of Sunshine: Gary Mormino, *Land of Sunshine, State of Dreams: A Social History of Modern Florida* (Gainesville: University Press of Florida, 2005); and interview by author, December 8, 2009.

20 *John Gorrie:* See http://www.knowsouthernhistory.net/Biographies/ John_Gorrie.

20 *Mormino's Colleague Raymond Arsenault:* interviews by author, St. Petersburg, FL, December 8, 2008.

20 *his celebrated 1984 article on air-conditioning:* Raymond Arsenault, "The End of the Long, Hot Summer."

20 *spends $35,000 to $40,000:* Scripps Southwest Florida Group, "Naples, Florida: The Market Difference," http://www.ndnadvertising.com/PDFs/ NDN_Media_Kit.pdf.

20 *cathedral of air-conditioned culture:* Raymond Arsenault, interview by Max Linsky, "The A/C Ph.D.," Tampa *Creative Loafing,* June 20, 2007.

21 *vehicle's air conditioner consumes seventy-three gallons of gasoline:* National Renewable Energy Laboratory, 2009, http://www.nrel.gov/vehicle sandfuels/ancillary_loads/ac_emissions.html.

21 *neither home builders nor air-conditioning contractors could recall:* Rick Porter, "A/C Nation," Port Charlotte *Sun-Herald,* June 10, 2000, www .sun-herald.com/2000/fron17.htm.

22 *There's no moral issue here:* Melanie Ave, "For Them, No A/C Is Really No Problem," *St. Petersburg Times,* July 22, 2006.

22 *Half of all mobile homes:* Mormino, *Land of Sunshine, State of Dreams.*

22 *trailers are unfit for human habitation:* Ashley Depenbusch, Student/

Farmworker Alliance, for Coalition of Immokalee Workers, telephone interview by author, December 5, 2008.

23 *River of Grass:* The name was bestowed by Marjory Stoneman Douglas, *The Everglades: River of Grass* (Sarasota, FL: Pineapple Press, 2007).

23 *An in-depth analysis of satellite imagery:* Julie Hauserman et al., "Florida's Coastal and Ocean Future: A Blueprint for Economic and Environmental Leadership," Issue Paper, Natural Resources Defense Council, September 2006, http://www.nrdc.org/water/oceans/florida/flfuture .pdf.

24 *South Florida leads the nation in water consumption per person:* Grunwald, *The Swamp,* 362.

24 *While farmers use far more water:* Cynthia Barnett, *Mirage: Florida and the Vanishing Water of the Eastern United States* (Ann Arbor: University of Michigan Press, 2007), 76.

24 *Environment Florida found:* Alison Cassady and Emily Figdor, *Feeling the Heat: Global Warming and Rising Temperatures in the United States* (Tallahassee, FL: Environment Florida Research and Policy Center, 2007).

24 *Human-caused changes:* R.A. Pielke et al., "The Influence of Anthropogenic Landscape Changes on Weather in South Florida," *Monthly Weather Review* 127 (1999): 1663–1673.

25 *rising seas could flood 15 percent:* Elizabeth Stanton and Frank Ackerman, "Florida and Climate Change: The Costs of Inaction," *Environmental Defense,* November 2007, http://ase.tufts.edu/gdae/Pubs/rp/FloridaClimate .html.

25 *There's no power on earth that can stop it!:* Michael Grunwald, "Growing Pains in Southwest Florida: More Development Pushes Everglades to the Edge," *Washington Post,* June 25, 2002.

25 *Carl Icahn:* Laura Layden, "Icahn Unloads WCI Stock for 2 Cents," *Naples Daily News,* December 9, 2008.

25 *As Detroit must sell cars:* Rothchild, *Up for Grabs,* 95.

26 *Nicole Ryan:* telephone interview by author, December 9, 2008.

26 *Ellen Peterson:* telephone interview by author, December 9 and 12, 2008.

26 *Napalm Beach:* Grunwald, *The Swamp,* 365.

26 *Patty Huff:* telephone interview by author, December 10, 2008.

27 *Rita Parker:* telephone interview by author, December 10, 2008.

28 *acceptance of physical discomfort:* Rothchild, *Up for Grabs,* 105–6.

28 *Think about it:* Gary Mormino, interview by author, December 8, 2008.

29 *Following a tornado that killed twenty residents:* David Damron and

Stephen Hudak, "Will You Consider Tornado Sirens?" *Orlando Sentinel,* February 6, 2007.

29 *the price of air-conditioning had dropped:* Arsenault, "End of the Long, Hot Summer"

29 *Florida, like Arizona, had stopped growing:* Ron Word, "Florida Population Drops for First Time Since 1946," *Naples Daily News,* August 17, 2009; Elysa Batista, "Bureau Estimate Shows Collier Population Increased, While Many Left Naples," *Naples Daily News,* August 19, 2009; Haya El Nasser, "For Florida, 'End of an Era' of Population Growth," *USA Today,* September 1, 2009.

Chapter 2

31 *Air-conditioners in office buildings:* Helen Meredith, "As Our Summers Get Warmer, The Race Is On to Design More Efficient Buildings and Cooling Systems," *The Age,* May 6, 2008.

31 *It is your human environment:* Mark Twain, *Following the Equator: A Journey Around the World* (Hartford, CT: American Publishing Company, 1897), 109. Twain is also credited, somewhat uncertainly, with that most well-known of all comments on the subject: "Everybody talks about the weather, but nobody does anything about it." Today, humanity appears to be negating the last half of that statement.

31 *On hot summer days:* Agis Papadopoulos, "The Influence of Street Canyons on the Cooling Loads of Buildings and the Performance of Air-Conditioning Systems," *Energy and Buildings* 33 (2001), 601–07. See also M. Beccali, M. Cellura, V. Lo Brano, and A. Marvuglia, "Short-Term Prediction of Household Electricity Consumption: Assessing Weather Sensitivity in a Mediterranean Area," *Renewable and Sustainable Energy Reviews* 12 (2008), 2040–65.

32 *analysts at Hamburg University of Technology:* Britta Stein and Christina Rullán Lemke, "Population—Climate—Energy: Scenarios to 2050," Fifty-first International Federation for Housing and Planning Congress, Copenhagen, September 23–26, 2007, http://www.klima2008.net/index .php?a1=pap&cat=1&e=59.

32 *air-conditioning is approaching 20 percent:* In 2007, U.S. residences consumed 261 billion kilowatt-hours (kWh) for air-conditioning out of a total 1,392 billion kWh, an 18.8 percent share and growing. The commercial sector used 173 billion kWh for air-conditioning in 2007, out of a total of 1,336 billion. Manufacturing used approximately 50 billion kWh for comfort air-conditioning in 2002, out of 966 billion total. I did

not include approximately 63 billion kWh used for "process" air-conditioning, which is done as part of manufacturing processes, not primarily to provide comfort. Data for all three sectors are from surveys by the federal Energy Information Administration (EIA), part of the Department of Energy. Residential and commercial data for 2007 and future years are from EIA's "Annual Energy Outlook 2009," http://www.eia.doe.gov/oiaf/aeo. More detailed data from the most recent surveys for all three sectors, as well as past surveys, may be found at EIA, "Households, Buildings, Industry and Vehicles," 2009, http://www.eia.doe.gov/emeu/consumption/index.html.

32 *a total of 497 billion kilowatt-hours for all uses:* EIA, "Annual Energy Review 2007," Table 8.9, http://www.eia.doe.gov/emeu/aer/elect.html.

32 *We use as much:* EIA, "International Electricity Consumption," undated, http://www.eia.doe.gov/emeu/international/electricityconsumption.html.

32 *Imagine that:* Ibid. I multiplied U.S. per capita consumption for air-conditioning by the total of the most recent population estimates for India, Indonesia, and Brazil and compared that figure with total consumption for all purposes by those three countries plus Mexico, Italy, the UK, and the continent of Africa.

33 *circumstances of society's own making:* See chapter 5 of this book and Eric Klinenberg, *Heat Wave: A Social Autopsy of Disaster in Chicago* (Chicago: University of Chicago Press, 2003).

34 *invention is, in turn the mother of necessity:* The context in which Veblen said this isn't entirely clear, but he is typically credited with it. Plato wrote in *Republic,* Book II, "the true creator is necessity who is the mother of invention."

34 *Table 1:* Data are based on *ASHRAE Fundamentals Guide* (Atlanta: American Society of Heating, Refrigeration, and Air-Conditioning Engineers, 1985), 26.21.

34 *The function of a steam engine:* This and the preceding discussion are based on author's interviews of Raymond Dean, professor emeritus of electrical engineering and computer science at the University of Kansas. Conversion of heat into mechanical power depends on the absolute temperatures, T_{HIGH} and T_{LOW}, of the heat source (where the heat is coming from) and heat sink (where we want it to go), respectively. Specifically, the maximum possible conversion efficiency is given by the formula:

Maximum Heat-Engine Efficiency = $(T_{HIGH} - T_{LOW}) / T_{HIGH})$

Thus, for a given heat flowing into the engine, the amount of mechanical power output is directly proportional to the difference between the temperature of the heat source and the temperature of the heat sink. This formula shows that heat-engine efficiencies are necessarily less than 100 percent—often much less. For example, the efficiency of a large gas turbine is around 30 percent, and the efficiency of a typical car engine is around 15 percent. But what does this tell us about refrigeration? The second law of thermodynamics applies equally well if the heat machine runs "backward": in a heat engine, heat flowing "downhill" from higher to lower temperature produces mechanical power output. Conversely, in a heat pump, mechanical power is the input, not the output. It forces heat to flow "uphill" from lower to higher temperature. How much mechanical power does it take to move heat against its natural direction of flow? Again, it depends on the temperatures, T_{HIGH} and T_{LOW}. Specifically, the maximum possible efficiency (maximum heat out divided by mechanical power in), is given by the formula:

Maximum Heat-Pump Efficiency = $T_{HIGH} / (T_{HIGH} - T_{LOW})$

Notice that the maximum efficiency for a heat pump is the maximum efficiency for a heat engine, flipped upside down. That should be no surprise because a heat pump is just a heat engine running backward. Many will recognize the term "heat pump" as a machine that uses electricity to heat a house in winter by moving heat from outdoors where it's cold to indoors where it's warm. (The temperature may be low outside, but it's far above absolute zero—the theoretical temperature, –460°, at which heat is absent—so there is still plenty of heat that can be squeezed out of the outdoor air.) In such a heat pump, the machine supplies heat at high temperature. But more commonly, we want to *remove* heat at *low* temperature. Familiar examples of this type of heat pumping are refrigerators and air conditioners. In this case we care more about heat inflow than heat outflow, and the maximum possible efficiency (maximum heat in divided by mechanical power in), is given by the formula:

Maximum Air-Conditioner Efficiency = $T_{LOW} / (T_{HIGH} - T_{LOW})$

For both applications, $(T_{HIGH} - T_{LOW})$ is usually much smaller than either T_{HIGH} or T_{LOW}. Therefore, efficiencies of heat pumps, refrigerators, and air conditioners are typically significantly greater than 100 percent, with values like 300 percent to 500 percent. That doesn't mean they are violating the laws of thermodynamics by producing new energy. It simply means that the quantity of existing heat energy they can move from one place to another is greater than the energy required to move it. This apparent energy-multiplication effect is an important economic factor, but it does not take into account the much larger quantity of energy used up by the power plants and transmission lines that supply electricity to the house. The second law of thermodynamics is always in effect: energy is not actually being multiplied.

35 *Figure 1:* The figure was drawn with the assistance of Raymond Dean.

35 *Figure 1:* Chilled, circulating refrigerant—a material that changes from liquid to gas and back at the right temperatures—absorbs heat from the warmer air in a house. After the indoor air cools down to the dew point, further cooling begins to wring out water, and the amount of water vapor that the air can hold continues to decrease as the temperature of the air continues to fall. The refrigerant takes on all of that heat and carries it out of the house. Entering the guts of the air conditioner—the compressor— the refrigerant is squeezed into a smaller volume. That raises its temperature well above that of the outside air, because as the pressure of a gas increases, so does its temperature. In this process, the electrical energy used to run the compressor is added to the heat energy which the refrigerant has just absorbed from the house. As it is pumped through the coils of the unit that sits outside the house, the refrigerant's temperature is higher than that of the outdoor air, even on the hottest day. By condensing into a liquid as it flows through the outdoor heat exchanger (called the condenser), the refrigerant loses the heat that it picked up from the house's interior, as well as the energy that was used to drive the compressor. Back inside the house, and immediately before it enters the indoor heat exchanger, the liquid refrigerant's pressure is rapidly reduced as it flows through a controlled opening called the expansion valve. As its pressure decreases, a small amount of evaporation makes the refrigerant's temperature plummet. Then, this well-chilled fluid—which is still mostly liquid—is ready to absorb more heat by completely evaporating into a gas as it flows through the indoor heat exchanger, which is called the evaporator.

There are many variations on this theme. In a window-style air conditioner, the components shown in this diagram are contained within a single box. The "split" room air conditioner popular in other countries is arranged very much as shown in the diagram. A typical residential central air system does not simply blow cool air into the house as shown, but circulates cool air throughout the house via ductwork. Many large buildings also circulate cool air from a main unit (which often is located on the roof) through ducts. Others have a large central unit that chills water and circulates it to smaller heat-exchanging units throughout the building. Those indoor units blow air over coils through which the chilled water circulates. Or each room might have its own compressor that circulates refrigerant within the room's individual unit; heat absorbed by the refrigerant is transferred to water that circulates through the building and to an outdoor cooling tower, where it sheds the heat.

The *power* going into and out of the pictured system is expressed in kilowatts. In the everyday world of heating and air conditioning, a common unit of heat *energy* is the British thermal unit (Btu). For quantities of energy in the form of electricity, the watt-hour is the usual basic unit. One watt-hour is equivalent to a little more than 3.4 Btu. Power—the rate of "flow" of energy at a given point in time—can be expressed in Btu per hour or as watts. The rate at which an air conditioner can remove heat from a space can be expressed in Btu per hour, or, more commonly, in tons, one ton being equivalent to 12,000 Btu per hour. (It's called a "ton" because it is the amount of power required to melt 2,000 pounds of ice in 24 hours.) The *quantity* of electricity consumed by an appliance, a home, or a society in a year, for example, might be expressed in kilowatt-hours or megawatt-hours (i.e., thousands or millions of watt-hours), whereas the maximum *rates* at which power plants supply energy or, say, the *power* required to run an air conditioner at a given setting would be expressed in megawatts or kilowatts.

One more thing: We often see efficiency of air conditioners and other appliances expressed as energy efficiency ratio (EER). EER is defined as Btu per hour of cooling power divided by the number of watts of electrical power required at a given setting. For air conditioning, another common rating, the seasonal energy efficiency ratio (SEER), is considered more useful to the consumer; it is computed as the total Btu of cooling output over an entire cooling season divided by the total energy (in watt-hours) consumed. The SEER spans days when air-conditioning systems have to run flat-out, along with days that require the unit to run less or not at all. Older central air-conditioning units for homes and window air conditioners might have SEER ratings around ten, whereas central air-conditioning systems sold since the start of 2006 in the United States must have SEER ratings of thirteen or higher. Conventional air conditioners and heat pumps with SEER of twenty or higher are available but not widely used because of high initial cost.

36 *seven times as much energy:* EIA, "Households, Buildings, Industry and Vehicles."

36 *Back in 1960:* Ibid., and Arsenault, "The End of the Long, Hot Summer."

37 *Commercial and public buildings:* Computed from EIA's 1995 and 2003 "Commercial Buildings Energy and Consumption Survey, http://www.eia.doe.gov/emeu/cbecs/contents.html. Top users of energy for air-conditioning in 2003 were retail (32 billion kWh), offices (30), education (22), and health care (10).

37 *Vacancy rates in the commercial sector rose:* Rates increased to 11 percent in retail and 12 percent for offices in 2008–09. See Cari Brokamp, "Retail Hits Highest Vacancy Rate in a Decade," GlobeSt.com, May 19, 2009, http://www.globest.com/news/1412_1412/chicago/178729-1.html; and "U.S. Office Vacancy Rate Hits Three-Year High," *Real Estate Weekly*, April 22, 2009.

37 *the government expects consumption for cooling overall to leap:* Energy Information Agency, "Annual Energy Outlook, 2009."

37 *By 2007, summertime peak:* EIA, *Electric Power Annual 2007*, DOE/EIA-0348, January, 2009, http://www.eia.doe.gov/cneaf/electricity/epa/epa.pdf.

38 *Our electrical infrastructure:* Gene Wolf, "Power by the Hour," *Transmission and Distribution World*, April 1, 2008.

38 *Close to half:* Janice Houston, Sara Sanchez, and Richard Pak, "Peak Load Growth Along the Wasatch Front: What's Driving Electricity Demand in Utah?" Utah Foundation, Report Number 663, January 2004.

38 *In Greece:* S. Hassid et al., "The Effect of the Athens Heat Island on Air-Conditioning Load," *Energy and Buildings* 32 (2000), 131–41.

38 *doubled its capacity to generate:* EIA, "Annual Energy Review 2007."

38 *a permanent decline in world natural gas supplies:* See, for example, Julian Darley, *High Noon for Natural Gas* (White River Junction, VT: Chelsea Green, 2004).

39 *gas reserves had grown to more than 2,000 trillion cubic feet:* Jad Mouawad, "Estimate Places Natural Gas Reserves 35% Higher," *New York Times*, June 17, 2009.

39 *the Marcellus Basin:* Tom Gjelten, "Rediscovering Natural Gas by Hitting Rock Bottom," *Morning Edition*, National Public Radio, September 22, 2009, http://www.npr.org/templates/story/story.php?storyId=113043935&ps=rs.

39 *55 percent of the level:* I divided total carbon emissions by total energy produced in 2008 at all U.S. natural-gas power plants and computed the same ratio for coal plants. Emissions per output for gas were 55 percent as high as for coal. Data from EIA, "Annual Energy Outlook 2009."

39 *almost 5 million metric tons of methane:* EPA, "2009 U.S. Greenhouse Gas

Inventory Report," April 2009, http://epa.gov/climatechange/emissions/
usinventoryreport.html.

40 *federal, state, and private investigators:* Abrahm Lustgarten, "Does Natural
Gas Drilling Endanger Water Supplies?" *BusinessWeek*, November 11,
2008; New York State Department of Environmental Conservation,
"Draft Supplemental Generic Environmental Impact Statement On The
Oil, Gas and Solution Mining Regulatory Program," September 2009,
ftp://ftp.dec.state.ny.us/dmn/download/OGdSGEISFull.pdf; Endocrine
Disruption Exchange, "Drilling and Fracturing Chemicals," 2009, http://
www.endocrinedisruption.com/chemicals.fracturing.php.

40 *Drilling in shale can consume hundreds:* Joaquin Sapien, "With Natural
Gas Drilling Boom, Pennsylvania Faces an Onslaught of Wastewater,"
ProPublica, October 3, 2009.

41 *up to 267 times:* Abrahm Lustgarten, "Is New York's Marcellus Shale Too
Hot to Handle?" *ProPublica*, November 9, 2009.

41 *T. Boone Pickens:* Pickens's plan is outlined at http://www.pickensplan
.com/theplan.

41 *major environmental groups:* See, for example, Natural Resources Defense
Council, "Finding the Balance: The Role of Natural Gas in America's Energy Future," September 30, 2008, http://www.nrdc.org/energy/balance
.asp. In an effort to defuse March 3, 2009, climate protests and civil disobedience aimed at the Congress and its coal-burning Capitol Power
Plant in Washington, House speaker Nancy Pelosi announced on February 26 that the plant would be converted to run on natural gas. Writer Bill
McKibben, one of the protest leaders, reportedly emailed friends, "just to
say, this civil disobedience stuff kind of works. How many coal plants are
there?" See Joseph Romm, "Pelosi and Reid: No More Coal for Capitol
Power Plant," Climate Progress, February 26, 2009, http://climateprogress
.org/2009/02/26/pelosi-and-reid-no-more-coal-for-capitol-power-plant.
Certainly, coal must be phased out as quickly as possible, but a gas-
drilling boom won't accomplish that. From 1996 to 2007, as electric utilities' gas consumption doubled, their coal use continued to climb steadily.
But what if we really do get serious about retiring coal plants? Suppose
for example that with the shale-drilling bonanza, we can achieve yet another doubling of gas-fired electricity generation, but this time eliminate
an equivalent amount of coal-fired generation. Even that steep escalation
of gas drilling would cut the utility industry's carbon emissions by only
12 percent and the nation's total carbon emissions by just 5 percent. (Calculations are based on 2008 emissions figures in EIA, "Annual Energy
Outlook, 2009".) Likewise, substituting natural gas for gasoline in all ve-

hicles as suggested in T. Boone Pickens's plan (see previous note) would reduce the nation's total carbon emissions by less than 9 percent, if we accept Pickens's claim that vehicle emissions would be reduced by 25 percent. Worse, converting all gasoline-powered vehicles would consume more natural gas than is currently consumed by electric utilities, homes and businesses combined (see EIA, "Natural Gas Consumption by End Use," 2009, http://tonto.eia.doe.gov/dnav/ng/ng_cons_sum_dcu_nus_a .htm). That increased gas demand would have disastrous consequences for the nation's water supplies. The only real solution to the emissions problem is a deep cut in energy consumption.

41 *The air-conditioning of buildings in America:* Add up the electrical energy used to run air-conditioning systems in residential, commercial, and industrial buildings (but not in vehicles), take into account the fuels used to generate the national electricity supply, apply the Environmental Protection Agency's national average of 1.37 pounds of carbon dioxide emissions per kilowatt-hour delivered, and you get 331 million tons. Alternatively, add together the residential and commercial cooling emissions reported in EIA, "Annual Energy Outlook 2009," leaving out manufacturing-sector emissions entirely, and the total comes to 305 million tons (after conversion from the metric tons reported by EIA). The average U.S. passenger car travels 12,500 miles and emits 10,000 pounds of carbon dioxide per year, according to the EPA (http://www.p2pays. org/ref/20/19353.pdf). If 110 million households each drove an additional car 7,000 miles, emitting 5,600 pounds of carbon dioxide each, the total comes to 308 million tons.

41 *To shoulder a fair share:* See http://unfccc.int/2860.php; http://www .hm-treasury.gov.uk/sternreview_index.htm; George Monbiot, *Heat: How to Stop the Planet from Burning* (Cambridge, MA: South End Press, 2007); and Francois Marechal, Daniel Favrat, and Eberhard Jochem, "Energy in the Perspective of the Sustainable Development: The 2000 W Society Challenge," *Resources, Conservation and Recycling* 44 (2005), 245–62.

42 *the job will be more manageable in a lower-energy society:* But some energy sources make us happier than do others. In public discussion of energy issues, the internal combustion engines that power our vehicles tend to generate the most noise and get the most attention. But aside from taking us places, look at what driving does for us: the misery of traffic jams, fume-filled nostrils, road rage, the constant threat of injury or death, extra expenses, and noise, noise, noise. As car buyers, we're motivated by thoughts of freedom and fun, but the car ends up, more often than not, taking us to places where we really don't want to go. For those who do

manage to travel on foot, traffic creates even more inconvenience and no shortage of danger. I believe that we Americans have an unconscious attachment to electricity that is far stronger than our widely acknowledged (and often seemingly irrational) attachment to gasoline. Electricity is like an old companion who's always standing by to lend a hand (given that you follow standard precautions against electrocution; and for the sake of mental calm, it's best not to think about the mining and burning that happens somewhere else to produce it). Electric power shapes America's home life more than any other factor, and does so silently and unobtrusively, aside from some occasional carrot chopping or lumber sawing or commercial broadcasting. Its benefits are fully appreciated only on those rare occasions when it fails. Even then, we take it for granted; think about how many times during a power outage you've absentmindedly thrown on a switch.

42 *Just under 50 percent:* EIA, "Annual Energy Outlook 2009."

43 *Electrical generation from renewable sources:* Electric generation from renewable sources, excluding hydroelectric, was 97 billion kWh in 2007, projected to grow to 429 billion kWh by 2030, a year in which air-conditioning demand is projected to reach 574 billion kWh. Data are from EIA, "Annual Energy Outlook 2009."

43 *If all household vehicles:* According to the Federal Highway Administration (http://www.fhwa.dot.gov/policy/ohpi/hss/hsspubs.cfm), there were a total of 2.8 trillion vehicle-miles traveled by passenger cars and light trucks in 2007. Using a typical energy efficiency of 3 miles per kilowatt-hour (see, e.g., the Idaho National Laboratory, http://avt.inl.gov), we find that traveling the same number of miles in electric vehicles using current technology would consume 926 billion kWh per year, almost exactly twice as much electricity as is used in the residential and commercial sectors. That ratio would be reduced with use of more energy-efficient electric vehicles. The ratio would be larger if sport-utility vehicles, vans, and pickup trucks were replaced by similarly sized electric vehicles.

44 *People are increasing their comfort expectations:* Yolande Strengers, "Comfort Expectations: The Impact of Demand-Management Strategies in Australia," *Building Research and Information* 36 (2008), 381–91.

44 *Chinese households:* Simon Wang, "China Needs Air-Conditioning," *Forbes,* February 10, 2008.

44 *I see AC sales competing:* Shailendra Bhatnaga, "Steamy India Discovers Joys of Air-Conditioning," Reuters, October 17, 2002.

44 *In Dubai:* "Dubai's Palazzo Versace to Have Air-Con Beach," *Arabian Business,* December 15, 2008.

44 *In the title of his 2000 essay collection:* Cherian George, *Singapore: The Air-Conditioned Nation* (Singapore: Landmark Books, 2000); and http://www.eskibar.com.

44 *a canine "hotel":* Wasti Atmodjo, "Air-Conditioned Rooms for Rent, for Dogs," *Jakarta Post,* April 10, 2008.

45 *In July 2005, South West Trains:* Ben Webster, "Sorry, It's Too Hot for Air-Conditioning, Rail Travellers Told," *The Times* (London), July 19, 2005.

45 *Maricopa County Jail:* Ananda Shorey, "Phoenix Is Sizzling Through What Could Be the Hottest July on Record," Associated Press, July 25, 2003; Maricopa County Sheriff's Office, "M.A.S.H," 2009, http://www.mcso.org/index.php?a=GetModule&mn=Mash.

45 *The United Nations Secretariat:* Betsy Pisik, "U.N. Thermostat to Be Set Higher," *Washington Times,* July 30, 2008.

46 *Prayas Energy Group:* Energy is one of four main areas in which Prayas works, the others being health, livelihoods, and education. Its members are, according to the group's Web site, "professionals working to protect and promote public interest in general and interests of disadvantaged sections of society in particular." See http://www.prayaspune.org.

46 *no one would dispute that India:* Ashok Sreenivas, Daljit Singh, and Girish Sant, "Climate Change: Separating the Wheat from the Chaff," *Economic and Political Weekly* 44, no. 5 (2009), 19–22.

46 *Greenhouse Development Rights Framework:* Tom Athanasiou, Sivan Kartha, and Paul Baer, "A Peak on the Horizon," Earth Island Institute, August 4, 2008, http://www.ecoequity.org/2008/08/a-peak-on-the-horizon.

47 *Beijing expanded a rural program:* Wu Qi, "China Boosts Rural Consumption with Household Appliance Subsidy Program," Xinhua News Agency, January 2, 2009; "China Details Subsidy Policy on Home Appliance Replacement," Xinhua, July 4, 2009.

48 *Thomas Midgley:* J.R. McNeill, *Something New Under the Sun: An Environmental History of the Twentieth-Century World* (New York: W.W. Norton, 2000), 111.

48 *rapidly increasing production of air conditioners:* Keith Bradsher, "The Price of Keeping Cool in Asia; Use of Air-Conditioning Refrigerant Is Widening the Hole in the Ozone Layer," *New York Times,* February 23, 2007.

48 *To replace the potent ozone destroyer CFC-12:* Environmental Protection Agency, "Mobile Air-Conditioning," 2009, http://www.epa.gov/cppd/mac.

49 *HFC-152a:* EPA, "HFC-152a Secondary Loop Vehicle A/C Systems," http://www.epa.gov/cppd/mac/152a/FINAL HFC 152a REPORT.pdf.

49 *By 2000, the world had "banked":* L. Palandre, A. Zoughaib, D. Clodic, and L. Kuijpers, "Estimation of the Worldwide Fleets of Refrigerant and Air-Conditioning Equipment in Order to Determine Forecasts of Refrigerant Emissions," Earth Technology Forum, Washington, DC, April 2003.

50 *If we vanish, millions of CFC and HCFC:* Weisman, *The World Without Us*, 205.

50 *replacing HCFCs and HFCs:* James Calm, "Emissions and Environmental Impacts from Air-Conditioning and Refrigeration Systems," *International Journal of Refrigeration* 25 (2002), 293–305.

Chapter 3

52 *On our half acre:* Phyllis McGinley, "Suburbia: Of Thee I Sing," *Harper's*, December 1949, 78–82.

52 *These days, I hear people say:* Christian Warren, telephone interview by author, February 24, 2009.

53 *suburbia of suburbia:* David Brooks, "Patio Man and the Sprawl People," *Weekly Standard*, August 12–19, 2002. The outdoor life features prominently in Brook's profile of Patio Man: "There he is atop the uppermost tier of his multi-level backyard patio/outdoor recreation area posed like an admiral on the deck of his destroyer. In his mind's eye he can see himself coolly flipping the garlic and pepper T-bones on the front acreage of his new grill while carefully testing the citrus-tarragon trout filets that sizzle fragrantly in the rear.... The sun is shining. The people are friendly. The men are no more than 25 pounds overweight, which is the socially acceptable male paunch level in upwardly mobile America, and the children are well adjusted. It is a vision of the sort of domestic bliss that Patio Man has been shooting for all his life."

53 *the top ten influences:* Robert Fishman, "The American Metropolis at Century's End: Past and Future Influences," *Housing Policy Debate* 11, no. 1 (2000), 199–213.

54 *When the National Academy of Engineering chose:* George Constable and Bob Somerville, *A Century of Innovation: Twenty Engineering Achievements That Transformed Our Lives* (Washington, DC: Joseph Henry Press, 2003).

54 *A comprehensive statistical analysis:* Jeff Biddle, "Explaining the Spread of

Residential Air-Conditioning, 1955–1980," *Explorations in Economic History* 45 (2008), 402–23.

55 *the "California garden":* Gail Cooper, "Escaping the House: Comfort and the California Garden," *Building Research and Information* 36 (2008), 373–80.

56 *Anxious to incorporate air-conditioning:* Gail Cooper, *Air-Conditioning America: Engineers and the Controlled Environment, 1900–1960* (Baltimore: Johns Hopkins University Press, 1998), 152.

56 *Architects, builders and bankers:* Ibid., 142.

56 *In 1957, the Federal Housing Administration:* Ibid.; Arsenault, "The End of the Long, Hot Summer."

57 *the average square footage:* U.S. Census Bureau, "Characteristics of New Housing," 2009, http://www.census.gov/const/www/charindex.html# singlecomplete; A. Wilson and J. Boehland, "Small Is Beautiful: U.S. House Size, Resource Use, and the Environment," *Journal of Industrial Ecology* 9 (2005), 277–87.

57 *Fewer than half of American single-family houses:* U.S. Bureau of the Census, "Presence of Air-Conditioning in New One-Family Houses Completed," 2009, www.census.gov/const/C25Ann/sftotalac.pdf.

57 *On a 100° summer day in Kansas:* Analysis by Raymond Dean, interview by author, August 9, 2009.

57 *myriad other reasons:* On the causes and consequences of the big-house craze, see Stan Cox, "Size Matters: Bigger Is Badder," *Hartford Courant,* October 14, 2007.

57 *A surprising 17 percent:* National Association of Realtors, "Second-Home Owner Survey Shows Solid Market, Appetite for More," press release, May 11, 2006; "Study: 744,000 Homeless in United States," Associated Press, January 10, 2007. I assumed about two occupants per vacation home.

58 *It has been estimated that 46 million:* J. Levy, Y. Nishioka, and J. Spengler, "The Public Health Benefits of Insulation Retrofits in Existing Housing in the United States," *Environmental Health* 2, no. 4 (2003), doi: 10.1186/1476-069X-2-4.

59 *a big-yet-small $5 billion worth of economic stimulus:* Michael Cooper, "Stimulus Funds Spent to Keep Sun Belt Cool," *New York Times,* June 7, 2009.

60 *air-conditioning use in Japan:* Haruyuki Fujii and Loren Lutzenhiser, "Japanese Residential Air-Conditioning: Natural Cooling and Intelligent Systems," *Energy and Buildings* 18 (1992), 221–33.

60 *only about one-tenth of a house's total energy:* G.A. Keoleian, S. Blanchard,

and P. Reppe, "Life-Cycle Energy, Costs, and Strategies for Improving a Single-Family House," *Journal of Industrial Ecology* 4 (2000), 135–36.

60 *A 2005 study published in the* Journal of Industrial Ecology: Wilson and Boehland, "Small Is Beautiful." Footnotes in the original article define the terms used in this table: "Moderate houses have R-19 walls, R-30 ceilings, double-low-e (U = 0.36) vinyl windows, R-4.4 doors, infiltration of 0.50 ACH heating and 0.25 ACH cooling, and R-6 ducts in attic. R-value is a measure of resistance to heat flow; R-19 is comparable to RSI-3.3 in the metric system. 'Low-e' refers to low-emissivity coatings that allow sunlight through, but block long-wavelength heat radiation. 'ACH' refers to 'air changes per hour,' a measure of how tight the building envelope is. Poorly insulated house has R-13 walls, R-19 attic, insulated glass vinyl windows, R-2.1 doors, infiltration of 0.50 ACH heating and 0.25 cooling, and uninsulated ducts."

61 *According to recent reports from Australia:* Yolande Strengers, "Comfort Expectations: The Impact of Demand-Management Strategies in Australia," *Building Research and Information* 36 (2008), 381–91.

62 *McMansion communities:* J.L. Nasar, J.S. Evans-Cowley, and V. Mantero, "McMansions: The Extent and Regulation of Super-Sized Houses," *Journal of Urban Design*, 12 (2007), 339–58.

62 *discussion on AppraisersForum.com:* "Are Appraisers Part of the Problem?" *AppraisersForum*, January 4, 2006, http://appraisersforum.com/showthread.php?t=100975.

62 *erected by home owner associations:* Stan Cox, "The Property Cops: Homeowner Associations Ban Eco-Friendly Practices," *AlterNet*, April 26, 2007, http://www.alternet.org/environment/51001; and Stan Cox, "Homeowner Groups Often Eco-Unfriendly," *Atlanta Journal-Constitution*, August 6, 2008.

63 *A search of major newspaper archives:* Nasar et al., "McMansions."

63 *by a marginal sixty square feet per house:* Evans, "Size of New Homes Starts Shrinking."

63 *about 2,300 to 2,500 square feet in 2015:* Leslie Mann, "Bigger Isn't Always Better: Home Buyers Opting for Comfort, Quality over Size," *Chicago Tribune*, February 20, 2009.

64 *people are choosing a bigger box:* Jennifer Robison, "Less Space: The Next Frontier in Real Estate," *Las Vegas Review-Journal*, March 18, 2007.

64 *still far out of line with other nations, whether poor or affluent:* The United Nations says houses and apartments in Pakistan or Nicaragua typically provide one-third of a room per person; it's half a room per person in Syria and Azerbaijan, about one room in eastern Europe, an average of a

room and a half in western Europe, and two whole rooms per person in the United States and Canada (not counting spaces like bathrooms, hallways, porches, etc.). Furthermore, a "room" as defined by the UN—an area as small as 42 square feet—is not the kind of space that an American real-estate agent would be eager to walk potential buyers through. See data summarized in J.A. Rode, "Appliances for Whom?" *Personal and Ubiquitous Computing* 10 (2005), 90–94.

64 *vacancy rate rose from 1.3 to 1.4 percent:* U.S. Census Bureau, "Housing Vacancies and Homeownership," 2009, http://www.census.gov/hhes/www/housing/hvs/historic/index.html.

64 *Participation in sixteen different types of outdoor activities:* O.R.W. Pergams and P.A. Zaradic, "Evidence for a Fundamental and Pervasive Shift Away from Nature-Based Recreation," *Proceedings of the National Academy of Sciences* 105 (2008), 2295–300.

65 *a perfect summer day:* The quip has been attributed to James Dent by, among others, Rustin Dodd, "Summertime, and the Living Is Easy," *University Daily Kansan,* May 7, 2009, http://www.kansan.com/stories /2009/may/07/morning_brew.

65 *human interaction with the natural environment:* Arsenault, "The End of the Long, Hot Summer."

65 *cocooning of America:* Fishman, "The American Metropolis at Century's End."

65 *I am amazed:* telephone interview by author, February 2009.

66 *a host of other writers and researchers have expressed concern about our separation from nature:* See, for example, Edward Wilson, *Biophilia* (Cambridge, MA: Harvard University Press, 1986); Rachel Kaplan and Stephen Kaplan, *The Experience of Nature: A Psychological Perspective* (New York: Cambridge University Press, 1989); Mary Rivkin, *The Great Outdoors: Restoring Children's Right to Play Outdoors* (Washington, DC: National Association for the Education of Young Children, 1995); and Gary Nabhan and Stephen Trimble, *The Geography of Childhood: Why Children Need Wild Places* (Boston: Beacon Press, 1995).

66 *A kid today can tell you:* Richard Louv, *Last Child in the Woods: Saving Our Children from Nature-Deficit Disorder* (Chapel Hill, NC: Algonquin Books, 2005), 1.

66 *it's not even known precisely how much time:* Among the few studies attempting to quantify this is Sanford Gaster, "Urban Children's Access to their Neighborhood: Changes over Three Generations," *Environment and Behavior* 23 (1991), 70–85.

67 *Contact with nature:* Nancy Wells and Gary Evans, "Nearby Nature: A

Buffer of Life Stress Among Rural Children," *Environment and Behavior* 35 (2003), 311–30; other studies are cited in Louv, *Last Child in the Woods.*

67 *Even the most extensive research:* Louv, *Last Child in the Woods,* 109.

67 *Results of a national survey:* Romina Barros, Ellen Silver, and Ruth Stein, "School Recess and Group Classroom Behavior," *Pediatrics* 123 (2009), 431–36.

67 *Recess advocates also claim:* Walter Kirn and Wendy Cole, "What Ever Happened to Play?" *Time,* April 30, 2001.

68 *(NAYRE):* See http://www.nayre.org.

68 *Summer vacation once made good sense:* Frederick Hess, "Summer Vacation of Our Discontent," *Washington Post,* July 12, 2006.

68 *There is no time for leisure time:* Billee Bussard, "Flawed Research," 2006, http://www.summermatters.com/research.htm.

69 *After all, play needs to happen somewhere:* Kirn and Cole, "What Ever Happened to Play?"

69 *police were called to break up twelve fights:* Anna Prior, "Calling All Cars: Trouble at Chuck E. Cheese's, Again," *Wall Street Journal,* December 9, 2008.

70 *In Bakersfield, California, police are called:* Jorge Barrientos, "Cops Called to Kid-Friendly Venues More Than Strip Club," *Bakersfield Californian,* February 17, 2009.

70 *We really do need to get out more:* Quote is from David Nicholson-Lord, "Why We All Need to Get Out More," *New Statesman,* January 24 2005; also see David Nicholson-Lord, *Green Cities—And Why We Need Them* (London: New Economics Foundation, 2003).

Chapter 4

71 *The capital of the new planet:* Henry Miller, *The Air-Conditioned Nightmare* (New York: New Directions Publishing, 1945), 41.

71 *in a typical year between 1947 and 2005, 2.5 to 3.5 percent of the nation's residents moved:* U.S. Census Bureau, "Geographical Mobility/Migration," 2009, http://www.census.gov/population/www/socdemo/migrate .html.

72 *The scale of Sun Belt growth:* M. Lakshminarayan Sastry, "Estimating the Economic Impacts of Elderly Migration: An Input-Output Analysis," *Growth and Change* 23 (1992), 54–79.

72 *Table 3:* EIA, "Housing Characteristics 1993," http://www.eia.doe.gov/ emeu/recs/recs2g.html; EIA, "Consumption and Expenditures 1993," http://www.eia.doe.gov/emeu/recs/recs2c.html; EIA, "Households,

Buildings, Industry and Vehicles 2009," http://www.eia.doe.gov/emeu/consumption/index.html.

72 *White people were prominent:* Jeanne Biggar, "Reassessing Elderly Sunbelt Migration," *Research on Aging* 2 (1980), 177–90.

73 *multiplier effect:* Sastry, "Estimating the Economic Impacts of Elderly Migration."

74 *the majority of new arrivals in the Phoenix area:* Larsen and Harlan, "Desert Dreamscapes."

74 *sharpened a racial redistribution:* Stewart Tolnay, "The African American 'Great Migration' and Beyond," *Annual Review of Sociology* 29 (2003), 209–32.

74 *by 1960, 18 percent of Southern households:* Arsenault, "The End of the Long, Hot Summer."

74 *In the wake of the 2004 presidential election:* James Wiley, "Blame Air-Conditioning for Kerry Loss," *Casper Star-Tribune,* December 5, 2004.

75 *Gore would defeat:* See details in Stan Cox, "America's Air-Conditioned Nightmare," *AlterNet,* June 29, 2006, http://www.alternet.org/story/38154. Do the reverse computations and you get a mirror-image outcome. In 1960, John F. Kennedy beat Richard Nixon in the popular vote by a microscopic 0.17 percent but finished with a comfortable 84-point win in the Electoral College. Those electoral votes were based on the population distribution of 1950, when air-conditioning was rare and the southward migration was just beginning. Had they been distributed instead as they are today, Nixon would have captured the presidency in 1960. In the real world, he had to wait until after the 1968 election to move into the White House.

75 *This is regular Republican atmosphere:* Marsha Ackermann, *Cool Comfort,* 66.

76 *an average of 208,000 more domestic migrants:* Texas A&M University Real Estate Center, "California Population and Components of Change," 2008, http://recenter.tamu.edu/data/pops/pops06.htm.

76 *experiencing 153 percent more cooling degree-days:* I weighted the CDD and HDD experienced by an average resident of each state by that state's population and averaged the states within each relevant region. The CDD and HDD data are from National Climatic Data Center, "Monthly State, Regional, and National Cooling Degree Days Weighted by Population."

76 *used 52 percent more total energy per capita:* I totaled state-by-state data within regions, from EIA, "State Energy Data System," http://www.eia.doe.gov/emeu/states/_seds.html.

76 *Just as China and India now take care of a lot of America's industrial "dirty*

work": For just one example, in the realm of pharmaceuticals, see chapter 3 of Cox, *Sick Planet.*

77 *It's an interesting irony:* Raymond Arsenault, interviewed by author, St. Petersburg, FL, December 8, 2008.

77 *demographer William Frey:* William Frey, "Three Americas: The Rising Significance of Regions," *Journal of the American Planning Association* 68 (2002), 349–356.

78 *bursting of a 'migration bubble':* Hope Yen, "Economic Woes Slow U.S. Migration to Sun Belt Region," Associated Press, March 19, 2009.

78 *as late as 2007–08, ninety-four of the nation's hundred:* Frank Hobbs and Nicole Stoops, *Census 2000 Special Reports, Series CENSR-4, Demographic Trends in the 20th Century* (Washington, DC: U.S. Government Printing Office, 2002).

78 *commuters were spending an average of two and a half times:* David Schrank and Tim Lomax, "The 2007 Urban Mobility Report," Texas Transportation Institute (September 2007), http://mobility.tamu.edu.

78 *Table 4:* Ibid. and EPA, "Room Air Conditioner Calculator," 2009, http://www.energystar.gov/ia/business/bulk_purchasing/bpsavings_calc/CalculatorConsumerRoomAC.xls.

78 *home to one-third of the nation's motor vehicles and burn half of the fuel:* National Renewable Energy Laboratory, "Air-Conditioning and Emissions."

78 *Commuter Pain Index:* Janet Caldow, "Feeling the Pain: The Impact of Traffic Congestion on Commuters," IBM Corporation, May 30, 2008, http://www.ibm.com/press/attachments/IBM_Traffic_Congestion_WhitePaper.pdf.

80 *The earliest use of air-conditioning in transportation:* Ackermann, *Cool Comfort,* 57–61.

80 *The national share of commuters using public transportation:* U.S. Environmental Protection Agency Office of Transportation and Air Quality, "Greenhouse Gas Emissions from the U.S. Transportation Sector, 1990–2003," March 2006, http://www.epa.gov/oms/climate/420r06003.pdf.

80 *fourteen hours per year:* Schrank and Lomax, "2007 Urban Mobility Report."

80 *Automotive commuting:* Writer James Howard Kunstler is foremost among many who have condemned the suburban car culture as an ecological and cultural catastrophe, and in his 2006 book *The Long Emergency,* he showed why it will be painful, and maybe impossible, to back out of that ubiquitous American lifestyle when the end of the cheap petroleum arrives—as it will. See James Kunstler, *The Geography of No-*

where: The Rise and Decline of America's Man-Made Landscape (New York: Free Press, 1994); Kunstler, *The Long Emergency: Surviving the Converging Catastrophes of the Twentieth Century* (New York: Atlantic Monthly Press, 2006).

80 *review of the 2009 Lincoln Navigator: U.S. News & World Report,* undated, http://usnews.rankingsandreviews.com/cars-trucks/Lincoln_Navigator/ Interior/. The Navigator's suggested retail price is $53,120 to $59,020, and its fuel efficiency is 14 miles per gallon in city driving, 20 mpg on the highway.

81 *hostile environment:* Gwyn Prins noted, for example, that the phrase "human beings are born into a hostile environment" came not from an anthropological monograph but from page one of a textbook on air-conditioning engineering. Gwyn Prins, "On Condis and Coolth," *Energy and Buildings* 18 (1992), 251–58.

81 *These oversized vehicles:* Josh Lauer, "Driven to Extremes: Fear of Crime and the Rise of the Sport Utility Vehicle in the United States," *Crime, Media, Culture* 1 (2005), 149–168.

82 *About 5.5 percent:* D. Bharathan et al., "An Overview of Vehicle Test and Analysis from NREL's A/C Fuel Use Reduction Research," National Renewable Energy Laboratory Conference Paper NREL/CP-540-41155, June 2007. American car dealers do stock, or claim to stock, small vehicles without air-conditioning. They don't really expect to sell such cars, but their rock-bottom prices can be used in advertisements to lure thrifty customers into the showroom, where they will buy something perhaps a bit bigger, and always air-equipped.

82 *total at 10.6 billion gallons:* R. Farrington and J. Rugh, "Impact of Vehicle Air-Conditioning on Fuel Economy, Tailpipe Emissions, and Electric Vehicle Range," *Earth Technologies Forum,* Washington, DC, 30 October–1 November, 2000, http://www.nrel.gov/docs/fy00osti/28960.pdf.

82 *In Europe:* Graham Parkhurst and Richard Parnaby, "Growth in Mobile Air-Conditioning: A Socio-Technical Research Agenda," *Building Research and Information* 36 (2008), 351–62.

82 *guzzles 19 to 22 percent more:* Bharathan et al., "An Overview of Vehicle Test and Analysis"

83 *increase by 19 to 66 percent:* Farrington and Rugh, "Impact of Vehicle Air-Conditioning."

83 *global-warming potential of carbon dioxide:* EPA, "Mobile Air-Conditioning," 2009, http://www.epa.gov/cppd/mac.

83 *increased carbon monoxide and nitrogen oxides by 70 to 80 percent:* Farrington and Rugh, "Impact of Vehicle Air-Conditioning."

83 *In the long-running debate:* See, for example, Michael Austin, "Gas Pains: Mileage Myths and Misconceptions," *Car and Driver,* December 2008; Jim Motavalli, "The Air Out There: An Endless Windows-vs-Air-Conditioning Debate," *New York Times,* July 30, 2008; "Wind of Change," *New Scientist,* March 18, 2006; "Boom-Lift Catapult, A/C vs. Windows Down," *Mythbusters,* Discovery Channel, November 10, 2004.

83 *The total number of vehicle miles traveled in the United States has doubled:* According to Federal Highway Administration figures, passenger cars and light trucks (passenger vans, SUVs, and pickups) traveled a total of 2.8 trillion miles in 2007 (http://www.fhwa.dot.gov/policy/ohpi/hss/ hsspubs.cfm), compared with 1.4 trillion miles in 1990 (http://www .fhwa.dot.gov/ohim/summary95/vm201.pdf). The U.S. population increased by 22 percent in that period.

83 *In more than 85 percent:* The average commuting vehicle contains 1.14 people, which requires that a minimum of 85 percent of vehicles contains only one person each. Pat Hu and Timothy Reuscher, *Summary of Travel Trends: 2001 National Household Travel Survey* (Washington, DC: Federal Highway Administration, 2004), 31. Average occupancy figures are from Bureau of Transportation Statistics, "National Transportation Statistics," 2008, http://www.bts.gov/publications/national_transportation_ statistics.

84 *the California Air Resources Board:* Amy Littlefield, "California Will Have the Coolest Cars," *Los Angeles Times,* June 26, 2009.

84 *an astonishing 5 to 8 percent:* Michael Vandenbergh, Jack Barkenbus, and Jonathan Gilligan, "Individual Carbon Emissions: The Low-Hanging Fruit," *UCLA Law Review* 55 (2008), 1701–58.

85 *fast-food drive-through windows along a single five-mile stretch:* Wally Kennedy, "Convenience Consumes Gasoline," *Joplin Globe,* January 27, 2009.

Chapter 5

86 *General Electric has proved a more devastating:* Arsenault, "The End of the Long, Hot Summer."

86 *being too cold:* Amy Zipkin, "Some Like It Hot," *New York Times,* January 23, 2005.

86 *had an inbred habit of ignoring discomfort:* Ackermann, *Cool Comfort,* 54–55.

86 *high priest of climatic determinism:* A. Cash Koeniger, "Climate and Southern Distinctiveness," *Journal of Southern History* 54 (1988), 21–44.

87 *With the publication of his book:* Sydney F. Markham, *Climate and the Energy of Nations* (London: Oxford University Press, 1944).

87 *The inhabitant of Kansas:* Ibid., 177–78.

88 *a map in which he assigned:* Ibid., 193.

88 *its influence ceases:* Ibid., 205.

89 *[I]s it merely a coincidence:* Koeniger, "Climate and Southern Distinctiveness."

89 *After decades of false starts:* Arsenault, "The End of the Long, Hot Summer."

89 *In 1983, Mancur Olson:* Mancur Olson, "The South Will Fall Again: The South as Leader and Laggard in Economic Growth," *Southern Economic Journal* 49 (1983), 917–32.

90 *Figure 2:* For the two intervals, I calculated the growth rates by subtraction, using data from the Bureau of Economic Analysis, "Gross Domestic Product by State," 2009, http://www.bea.gov/regional/gsp/default.cfm.

91 *comparisons between rich temperate countries:* William Nordhaus, "Geography and Macroeconomics: New Data and New Findings," *Proceedings of the National Academy of Sciences* 103 (2006), 3510–3517.

92 *We also find evidence:* Melissa Dell, Benjamin Jones, and Benjamin Olken, "Climate Change and Economic Growth: Evidence from the Last Half-Century," National Bureau of Economic Research Working Paper No. 14132, June 2008, http://harrisschool.uchicago.edu/Programs/beyond/workshops/pol_econ_papers/fall08-jones.pdf.

92 *when saleswomen's feet swelled:* Ackermann, *Cool Comfort,* 54.

92 *Before air-conditioning:* "Stay Cool! Air-Conditioning America," National Building Museum, Washington, DC, May 1, 1999–January 31, 2000. Quoted text is at http://www.eweek.org/site/News/Features/staycool.shtml.

93 *stimulating the economy and ending the Depression:* Cooper, *Air-Conditioning America,* 118.

93 *War Production Board:* Cooper, *Air-Conditioning America,* 140–41.

93 *energy expended to cool the retail sector shot up by 66 percent:* Data from EIA, "Commercial Buildings Energy Consumption Survey 2007," http://www.eia.doe.gov/emeu/cbecs/contents.html.

93 *Interviewing young residents of Singapore:* Russell Hitchings and Shu Jun Lee, "Air-Conditioning and the Material Culture of Routine Human Encasement: The Case of Young People in Contemporary Singapore," *Journal of Material Culture* 13 (2008), 251–65.

94 *the more upscale retail establishments of Manhattan:* Allen Salkin, "Shivering for Luxury," *New York Times,* June 26, 2005.

94 *as President Richard Nixon used to do:* See, for example, Julie Nixon Eisenhower's reminiscences ("He had a fire going summer, winter. He loved just the simplicity.") in an August 11, 2001, interview with Larry King at http://transcripts.cnn.com/TRANSCRIPTS/0108/11/1klw.00.html.

94 *the amount of mall space:* EIA, "Commercial Buildings Energy Consumption Survey."

95 *Common effects of extreme heat:* Jerry Ramsey, Charles Burford, Mohamed Youssef Beshir, and Roger Jensen, "Effects of Workplace Thermal Conditions on Safe Work Behavior," *Journal of Safety Research* 14 (1983), 105–14.

95 *The story of a Columbia, South Carolina, plant:* Nancy Smith and Paul Lauritzen, "Summer Breezes—But Indoors," *Fabricating and Metalworking,* October 2001.

97 *at 1.8 miles per hour:* Christian Taber, applications engineer at Big Ass Fans, telephone interview by author, May 12, 2009.

97 *Thermal stratification:* Raymond Dean, interview by author, August 24, 2009.

97 *In its 2005 exposé of worker abuse in poultry processing:* Human Rights Watch, "Blood, Sweat, and Fear: Workers' Rights in U.S. Meat and Poultry Plants," January 24, 2005, http://www.hrw.org/reports/2005/usa0105/usa0105.pdf.

98 *A Wall Street stock trader:* Salkin, "Shivering for Luxury."

98 *installed, aptly enough, at the New York Stock Exchange:* Cooper, *Air-Conditioning America,* 14.

99 *By the reckoning of top officials at the Carrier Corporation:* Cooper, *Air-Conditioning America,* 160–62.

99 *David Wyon:* series of e-mail interviews with author, April 2009.

100 *In the last analysis, [a nation's] economy:* David Wyon, "The Effects of Indoor Air Quality on Performance and Productivity," *Indoor Air* 14 (Suppl. 7, 2004), 92–101.

100 *increase in productivity of 0.5% to 5%:* David Wyon, "Enhancing Productivity While Reducing Energy Use in Buildings," E-Vision 2000 Conference, Washington, DC, October 11–13, 2000, https://www.rand.org/pubs/conf_proceedings/CF170.1-1/CF170.1.wyon.pdf.

100 *sick building syndrome:* See chapter 6.

100 *a 5 to 13 percent productivity loss:* R. Kosonen and F. Tan, "The Effect of Perceived Indoor Air Quality on Productivity Loss," *Energy and Buildings* 36 (2004), 981–86.

100 *One trial, conducted in a real office:* L. Fang, D.P. Wyon, G. Clausen, and P.O. Fanger, "Impact of Indoor Air Temperature and Humidity in an Of-

fice on Perceived Air Quality, SBS Symptoms and Performance," *Indoor Air* 14 (2004), 74–81.

101 *Alan Hedge:* telephone and e-mail interviews by author, April 2009. Hedge's three studies had not been published as of the time of this writing; slides summarizing his Florida study may be seen at http://ergo .human.cornell.edu/Conferences/EECE_IEQandProductivity_ABBR .pdf

103 *Adam Korn and Andrew Keown:* Zipkin, "Some Like It Hot."

104 *female employees disproportionately occupy lower-level, nonmanagement jobs:* Government Accountability Office, "Women in Management: Analysis of Selected Data from the Current Population Survey," Report to Congressional Requesters, GAO-02-156, October 2001.

104 *found that 92 percent of the executive positions in the top 150 companies:* Cindy Krischer Goodman, "Fewer Women Winning Jobs in Executive Suites," *Miami Herald,* February 11, 2009.

104 *Americans were spending a greater number of hours per week at work:* Lonnie Golden and Helene Jorgensen, "Time After Time: Mandatory Overtime in the U.S. Economy," Economic Policy Institute Briefing Paper No. 120, January 1, 2002, http://www.epi.org/publications/entry/briefing papers_bp120/.

105 *In a Singapore call center:* K.W. Tham, "Effects of Temperature and Outdoor Air Supply Rate on the Performance of Call Center Operators in the Tropics," *Indoor Air* 14 (Suppl. 7, 2004), 119–25.

105 *A report from Thailand:* John Busch, "A Tale of Two Populations: Thermal Comfort in Air-Conditioned and Naturally Ventilated Offices in Thailand," *Energy and Buildings* 18 (1992), 235–49.

105 *Cool Biz:* Sebastian Moffett, "Japan Sweats It Out As It Wages War on Air-Conditioning," *Wall Street Journal,* September 18, 2007.

106 *A consumer-market analyst in Japan:* W. David Marx, "The Prisoner's Dilemma of Cool Biz," http://clast.diamondagency.jp/en/?p=54.

106 *the Gunbelt:* Ann Markusen, *The Rise of the Gunbelt: The Military Remapping of Industrial America* (New York: Oxford University Press, 1991).

107 *fully 85 percent goes to power air-conditioning:* Deborah Zaborenko, "U.S. Army Works to Cut Its Carbon 'Bootprint,'" Reuters, July 27, 2008.

107 *At best, the [Leopard's] air conditioner:* Stephen Priestley, "For Cooler Heads to Prevail, the CF/DND Is Moving Briskly: Climate Control Systems for Leopard Tanks in Afghanistan," *Canadian American Strategic Review,* January 2006.

108 *American troops:* Jon Grinspan, "The Jeep, the Humvee, and How War Has Changed," *American Heritage* Web site, August 1, 2007, http://www

.americanheritage.com/articles/web/20070801-jeep-humvee-wwII-iraq .shtml. However, the most controversial use of air-conditioning was as a tool in the interrogation of detainees. The *U.S. Army Field Manual,* last updated in 2006, forbids "inducing hypothermia." But a secret 2007 report by the International Committee of the Red Cross (ICRC), leaked in 2009, scrutinized the treatment of "high value detainees" held by the Central Intelligence Agency (CIA) in facilities at Guantanamo Bay, Cuba, and it shows that the CIA did not recognize the military's restrictions as valid. ICRC charged, "Detainees frequently reported that they were held for their initial months of detention in cells which were kept extremely cold, usually at the same time as being kept forcibly naked. The actual interrogation room was often reported to be kept cold." The ICRC report was published online (http://www.americanheritage.com/articles/web/ 20070801-jeep-humvee-wwII-iraq.shtml) by the *New York Review of Books* as an appendix to Mark Danner, "The Red Cross Torture Report: What It Means," April 30, 2009. Mohammed al-Qahtani was accused of having planned to serve as a twentieth 9/11 hijacker. He had been refused entry to the United States in August 2001 and was later captured in Afghanistan. This account of treatment he received is based on a late 2002– early 2003 army interrogation log from Guantanamo Bay: "For eleven days, beginning November 23, al-Qahtani was interrogated for twenty hours each day by interrogators working in shifts. He was kept awake with music, yelling, loud, white noise, or brief opportunities to stand. He was then subjected to eighty hours of nearly continuous interrogation until what was intended to be a 24-hour 'recuperation.' This recuperation was entirely occupied by a hospitalization for hypothermia that had resulted from deliberately abusive use of an air conditioner. Army investigators reported that al-Qahtani's body temperature had cooled to 95 to 97 degrees Fahrenheit and that his heart rate had slowed to thirty-five beats per minute. . . . The prisoner slept through most of the 42-hour hospitalization after which he was hooded, shackled, put on a litter and taken by ambulance to an interrogation room for twelve more days of interrogation." Steven Miles, "Medical Ethics and the Interrogation of Guantanamo 063," *American Journal of Bioethics* 7(2007), 5–11.

108 *If it comes to pass, the Army's planned shift:* Keith Johnson, "Army Green: U.S. Military Gunning to Curb Carbon 'Bootprint,'" *wsj.com,* July 28, 2008, http://blogs.wsj.com/environmentalcapital/2008/07/28

Chapter 6

109 *We don't use the air conditioner:* Loren Lutzenhiser, "A Question of Control: Alternative Patterns of Room Air-Conditioner Use," *Energy and Buildings* 18 (1992), 193–200.

109 *Eddie Slautas:* Bennie Currie, "Cooler Air Arrives, Death Toll Climbs," *USA Today,* August 2, 1999.

109 *one of 103 Chicagoans killed by the heat that week:* CDC, "Heat-Related Deaths—Chicago, Illinois, 1996–2001, and United States, 1979–1999," *Morbidity and Mortality Weekly Report* 52 (2003), 610–613.

109 *a Commonwealth Edison power cable:* Gretchen Ruethling, "Hundreds Evacuated in Chicago As Heat Wave Persists," *New York Times,* August 2, 2006.

110 *the U.S. Global Change Research Program:* Jennifer Dlouhy, "New Report Stresses Immediacy of Global Warming," *San Francisco Chronicle,* June 17, 2009; John Broder, "Government Study Warns of Climate Change Effects," *New York Times,* June 16, 2009.

110 *more than 550 city residents were killed by the record-breaking heat:* Klinenberg, *Heat Wave.*

111 *many people have also forgotten:* S.A. Chagnon, K.E. Kunkel, and B.C. Reinke, "Impacts and Responses to the 1995 Heat Wave: A Call to Action," *Bulletin of the American Meteorological Society* 77 (1996), 1497–506.

111 *air-conditioning has reduced fetal and infant mortality,:* Arsenault, "The End of the Long, Hot Summer."

111 *emergency rooms were overwhelmed with patients:* Richard Pérez-Peña, "The Blackout of 2003: Health Care; Early Panic Leads to a Rush on Emergency Rooms," *New York Times,* August 15, 2003.

111 *From 1949 to 1995, the frequency of heat waves:* D.J. Gaffen and R.J. Ross, "Increased Summertime Heat Stress in the U.S.," *Nature* 396 (1998), 529–30.

111 *extreme heat waves:* Dlouhy, "New Report Stresses Immediacy of Global Warming."

111 *killed 35,000 to 52,000 people in Europe:* S. Bhattacharya, "European Heat Wave Caused 35,000 Deaths," *New Scientist,* October 10, 2003. However, noted that report, "since reports are not yet available for all European countries, the total heat death toll for the continent is likely to be substantially larger." In 2006, some sources raised the tally to 52,000: Janet Larsen, "Setting the Record Straight: More than 52,000 Europeans Died from Heat in Summer 2003," Earth Policy Institute, July 28, 2006, http://www.earth-policy.org/index.php?/plan_b_updates/2006/update56.

112 *heat waves kill more than 1,800:* L.S. Kalkstein and J.S. Greene, "An Evaluation of Climate/Mortality Relationships in Large U.S. Cities and the Possible Impacts of a Climate Change," *Environmental Health Perspectives* 105 (1997), 84–93. EPA points out that results of all published studies "fall in a narrow range of roughly 1,700–1,800 total heat-attributable deaths per summer." That is much higher than the Centers for Disease Control and Prevention's (CDC) annual average of 182 deaths officially caused by "excessive heat due to weather conditions." The reason: published studies use statistical increases in deaths during heat waves, whereas the CDC tally depends on individual causes of death recorded by public officials (EPA, "Excessive Heat Events Guidebook," Publication No. EPA 430-B-06-005, June 2006).

112 *A 2003 study of twenty-eight U.S. cities:* R.E. Davis, P.C. Knappenberger, P.J. Michaels, and W.M. Novicoff, "Changing Heat-Related Mortality in the United States," *Environmental Health Perspectives* 111 (2003), 1712–18.

112 *reduced death rates by 42 percent:* Marie O'Neill, "Air-conditioning and Heat-Related Health Effects," *Applied Environmental Science and Public Health* 1 (2003), 9–12.

112 *Table 5:* Marie O'Neill, Antonella Zanobetti, and Joel Schwartz, "Disparities by Race in Heat-Related Mortality in Four U.S. Cities: The Role of Air-Conditioning Presence," *Journal of Urban Health* 82 (2005), 191–97.

113 *people who die because they neglect themselves:* Eric Klinenberg, "The Politics of Heat Waves: Victims of a Hot Climate and a Cold Society," *New York Times*, August 22, 2003. Klinenberg also mentioned this comment in his earlier book *Heat Wave*.

113 *People with central air:* L.G. Chestnut, W.S. Breffle, J.B. Smith, and L.S. Kalkstein, "Analysis of Differences in Hot-Weather-Related Mortality Across 44 U.S. Metropolitan Areas," *Environmental Science Policy* 1 (1998), 59–70.

113 *features of neighborhoods:* M.S. O'Neill, A. Zanobetti, and J. Schwartz, "Modifiers of the Temperature and Mortality Association in Seven U.S. Cities," *American Journal of Epidemiology* 157 (2003), 1074–82.

113 *Marie O'Neill:* Marie O'Neill, e-mail interview by author, July 5, 2009.

114 *Now you see air-conditioning pitched:* Christian Warren, telephone interview by author, February 24, 2009.

114 *If current greenhouse emissions continue:* EPA, "Excessive Heat Events Guidebook."

114 *Some of those taking refuge:* Sewell Chan and Jonathan Allen, "Across the City, It's (Finally) Too Darn Hot," *New York Times*, August 17, 2009.

115 *World Health Organization predicts:* World Health Organization, "Climate Change and Infectious Disease," 2009, http://www.who.int/global change/climate/summary/en/index5.html.

115 *At this time, scientists:* Centers for Disease Control and Prevention, "Vector-borne and Zoonotic Diseases," 2008, http://www.cdc.gov/ climatechange/effects/vectorborne.htm.

115 *Brownsville, Texas, and Matamoros, Mexico:* Joan Marie Brunkard et al., "Dengue Fever Seroprevalence and Risk Factors, Texas-Mexico Border, 2004," *Emerging Infectious Diseases* 13 (2007), 1477–83.

116 *Asian tiger mosquito:* CDC, "Information on *Aedes albopictus*," 2005, http://www.cdc.gov/ncidod/dvbid/Arbor/albopic_new.htm; "Ferocious Tiger Mosquito Invades the United States," ABC News, July 30, 2001, http://news.nationalgeographic.com/news/2001/07/0730_wiretigermoz .html.

116 *Paul Reitner:* Paul Reitner, "Climate Change and Mosquito-Borne Disease," *Environmental Health Perspectives* 109, Supplement 1 (2001), 141–61.

116 *two technologies, air-conditioning and television:* P.M. Gahlinger, W.C. Reeves, and M.M. Milby, "Air-Conditioning and Television As Protective Factors in Arboviral Encephalitis Risk," *American Journal of Tropical Medicine and Hygiene* 35 (1986), 601–10; "West Nile Virus Activity— United States, 2006," *Journal of the American Medical Association* 298 (2007), 619–21.

117 *a 2003 paper in the* Lancet: D. Menzies et al., "Effect of Ultraviolet Germicidal Lights Installed in Office Ventilation Systems on Workers' Health and Wellbeing: Double-Blind Multiple Crossover Trial," *Lancet* 362 (2003), 1785–91.

118 *A 2004 National Institutes of Health review:* D.M. Morens, G.K. Folkers, and A.S. Fauci, "The Challenge of Emerging and Re-Emerging Infectious Diseases," *Nature* 430 (2004), 242–49.

118 *After days or weeks in the heat:* J.P. McClung et al., "Exercise-Heat Acclimation in Humans Alters Baseline Levels and Ex Vivo Heat Inducibility of HSP72 and HSP90 in Peripheral Blood Mononuclear Cells," *American Journal of Physiology: Regulatory, Integrative and Comparative Physiology* 294 (2008), R185–191.

118 *heat-shock proteins:* Ibid.; Pope Moseley, "Heat Shock Proteins and Heat Adaptation of the Whole Organism," *Journal of Applied Physiology* 83 (1997), 1413–17; A. Tetievsky et al., "Physiological and Molecular Evidence of Heat Acclimation Memory: A Lesson from Thermal Responses and Ischemic Cross-Tolerance in the Heart," *Physiological Genomics* 34

(2008), 78–87; P.M. Yamada et al., "Effect of Heat Acclimation on Heat Shock Protein 72 and Interleukin-10 in Humans," *Journal of Applied Physiology* 103(2007), 1196–204.

118 *laboratory rats:* Tetievsky et al., "Physiological and Molecular Evidence."

119 *Research, much of it done by the U.S. military:* McClung et al., "Exercise-Heat Acclimation"; Paulette Yamada, e-mail interview by author, November 19, 2008.

119 *Michal Horowitz:* Michal Horowitz, e-mail interview by author, November 21, 2008.

120 *so vile that an attendant walked up and down:* Cooper, *Air-Conditioning America*, 83.

121 *industry had laid the philosophical foundations for its growth:* Cooper, *Air-Conditioning America*, 51–79.

121 *ventilation is inadequate in many classrooms:* J.M. Daisey, W.J. Angell, and M.G. Apte, "Indoor Air Quality, Ventilation and Health Symptoms in Schools: An Analysis of Existing Information," *Indoor Air* 13 (2003), 53–64.

121 *Studies in California, Brazil, France:* M.J. Mendell et al., "Elevated Symptom Prevalence Associated with Ventilation Type in Office Buildings," *Epidemiology* 7 (1996), 583–89; M.J. Mendell and A.H. Smith, "Consistent Pattern of Elevated Symptoms in Air-Conditioned Office Buildings: A Reanalysis of Epidemiologic Studies," *American Journal of Public Health* 80 (1990), 1193–99; G.S. Graudenz et al., "Association of Air-Conditioning with Respiratory Symptoms in Office Workers in Tropical Climate," *Indoor Air* 15 (2005), 62–66; P. Preziosi, S. Czernichow, P. Gehanno, and S. Hercberg, "Air-Conditioning at Workplace and Health Services Attendance in French Middle-Aged Women: A Prospective Cohort Study," *International Journal of Epidemiology* 33 (2004), 1120–23.

121 *Mark Mendell wrote:* Mark Mendell, "Air-Conditioning As a Risk for Increased Use of Health Services," *International Journal of Epidemiology* 33 (2004), 1123–26.

122 *respiratory illness is increased by 50 to 120 percent:* J.M. Samet and J.D. Spengler, "Indoor Environments and Health: Moving into the 21st Century," *America Journal of Public Health* 93 (2003), 1489–93.

123 *the rate of abdominal obesity:* C. Li, E.S. Ford, A.H. Mokdad, and S. Cook, "Recent Trends in Waist Circumference and Waist-Height Ratio Among U.S. Children and Adolescents," *Pediatrics* 118 (2006), e1390–98.

123 *in 2006, a team of twenty medical researchers:* S.W. Keith et al., "Putative Contributors to the Secular Increase in Obesity: Exploring the Roads Less Traveled," *International Journal of Obesity* 30 (2006), 1585–94.

124 *Decades of evidence:* C.P. Herman, "Effects of Heat on Appetite," in *Nutritional Needs in Hot Environments: Applications for Military Personnel in Field Operations,* ed. B.M. Marriott (Washington, DC: National Academy Press, 1993), 187–214; Pawel Wargocki and David Wyon, "The Effects of Moderately Raised Classroom Temperatures and Classroom Ventilation Rate on the Performance of Schoolwork by Children (RP-1257)," *HVAC&R Research* 13 (2007), 193–220.

124 *Drs. Masako and Maiko Kobayashi:* M. Kobayashi and M. Kobayashi, "The Relationship Between Obesity and Seasonal Variation in Body Weight among Elementary School Children in Tokyo," *Economics and Human Biology* 4 (2006), 253–61.

125 *vitamin D deficiencies:* Adit Ginde, Mark Liu, and Carlos Camargo, "Demographic Differences and Trends of Vitamin D Insufficiency in the U.S. Population, 1988–2004," *Archives of Internal Medicine* 169 (2009), 626–32.

125 *a strong link between:* Robert Scragg and Carlos Camargo Jr., "Frequency of Leisure-Time Physical Activity and Serum 25-Hydroxyvitamin D Levels in the U.S. Population: Results from the Third National Health and Nutrition Examination Survey," *American Journal of Epidemiology* 168 (2008), 577–86.

125 *with 64 percent of Americans:* National Sleep Foundation, "One-Third of Americans Lose Sleep Over Economy," press release, March 2, 2009.

126 *A study in Japan:* T. Ueno and T. Ohnaka, "Influence of Long-Term Exposure to an Air-Conditioned Environment on the Diurnal Cortisol Rhythm," *Journal of Physiological Anthropology* 25 (2006), 357–362.

126 *hygiene hypothesis:* Garry Hamilton, "Filthy Friends," *New Scientist,* April 16, 2005; E.W. Gelfand, 2003, "The Hygiene Hypothesis Revisited: Pros and Cons," http://www.medscape.com/viewarticle/452170. (The "cons" were in reference to specific mechanisms then being proposed; the author was not doubting that exposure to the outdoor environment has an effect.)

127 *Graham Rook:* Graham Rook and Christopher Lowry, "The Hygiene Hypothesis and Psychiatric Disorders," *Trends in Immunology* 29 (2008), 151–58; Graham Rook, "Review Series on Helminths, Immune Modulation and the Hygiene Hypothesis: The Broader Implications of the Hygiene Hypothesis," *Immunology* 126 (2009), 3–11. This is a very active research area. The latter article was part of a special issue of *Immunology* addressing the subject.

129 *A statistical summer "trough":* D.A. Lam and J.A. Miron, "The Effects of Temperature on Human Fertility," *Demography* 33 (1996): 291–305.

129 *has proven much stronger:* D.A. Seiver, "Trend and Variation in the Seasonality of U.S. Fertility, 1947–1976," *Demography* 22 (1985), 89–100.

129 *seasonality didn't decline until much later among America's nonwhite people:* D.A. Seiver, "Seasonality of Fertility: New Evidence," *Population and Environment* 10 (1989): 245–57.

Chapter 7

130 *Think about it: fifty-six million people displaced:* Arundhati Roy, *Power Politics* (Cambridge, MA: South End Press, 2001), 67–68.

130 *birds fell dead:* Omer Farooq, "India's Heat Wave Tragedy," BBC, May 17, 2002, http://news.bbc.co.uk/2/hi/south_asia/1994174.stm.

130 *Reverend J.M. Merk:* Quoted in Henry Blanford, *A Practical Guide to the Climates and Weather of India, Ceylon, and Burmah, and the Storms of Indian Seas* (London: Macmillan and Co., 1889), 127–29.

132 *the best attribute of an A/C:* Shailendra Bhatnaga, "Steamy India Discovers Joys of Air-Conditioning," Reuters, October 17, 2002.

132 *projected to grow almost tenfold compared to its 2005 level:* Michael McNeil et al., "Potential Benefits from Improved Energy Efficiency of Key Electrical Products: The Case of India," *Energy Policy* 36 (2008), 3467–76.

132 *Figure 3:* Data adapted from ibid.

133 *outages and brownouts are routine:* Prayas Energy Group, *Know Your Power: A Citizens' Primer on the Electricity Sector,* 2nd ed. (Pune, India: Prayas, 2006), 11–13; 2008 information from Girish Sant, Prayas, interview by author, January 8, 2009.

133 *1.2 trillion kilowatt hours in 2025:* Energy Information Administration, "India Energy Profile," http://tonto.eia.doe.gov/country/country_energy_data.cfm?fips=IN.

134 *displaced an estimated 350,000 people:* "Narmada: The Facts," *New Internationalist,* July 2001.

134 *two Sardar Sarovar projects each year!:* Prayas Energy Group, *Know Your Power.*

134 *One needs to ask, 'Why allow air-conditioning':* Girish Sant, telephone interview, January 2009.

134 *electric fans used by 77 percent:* Jonathan Ablett et al., "The 'Bird of Gold': The Rise of India's Consumer Market," McKinsey Global Institute, May 2007.

135 *A human pedaling:* Raymond Dean, e-mail interview by author, August 9, 2009.

135 *the air conditioner is far more humane than the electric fan:* The manufac-

ture of cold in the tropics without industrial fuels can be a labor-intensive proposition indeed. In 1872, one T.A. Wise reported in the journal *Nature* that ice was being produced outdoors in Bengal, "in large quantities" on the coolest nights of the year, even when temperatures hardly dipped below 50°. Work crews dug numerous 20-by-120-foot beds two feet deep, filled them with straw, and laid out thousands of small clay saucers on the straw, each saucer holding a few ounces of water. Wise calculated that "239 gallons of water were thus exposed to the night sky on each bed," and when the temperature fell low enough and the breeze was at precisely the right speed from the right direction (north-northwest), radiation of heat to the night sky cooled and froze the water's surface, sometimes freezing all the way to the bottoms of the saucers. Before sunrise, reported Wise with true colonial enthusiasm, "Upwards of two hundred and fifty persons, of all ages, are actively employed in securing the ice for some hours . . . , and this forms one of the most animated scenes to be witnessed in Bengal." T.A. Wise, "Ice-Making in the Tropics," *Nature* 5 (1872), 189–90.

135 *Darshan Bhatia:* interview by author, Hyderabad, India, January 10, 2001.

136 *The Punjab state government:* "Punjab Wants More Power to Tide Over Crisis," *Chandigarh Tribune*, July 26, 2004.

137 *Tamil Nadu:* "Power Cuts Add to Summer Woes," *Hindu*, April 29, 2006.

138 *eight times as much carbon into the air in 2020:* McNeil et al., "Potential Benefits from Improved Energy Efficiency."

138 *barely twenty thousand British civilians:* Denis Judd, *Empire: The British Imperial Experience from 1765 to the Present* (New York: Basic Books, 1998), 78–79.

139 *Stella:* Stella Mary (no last name), interview by author via translator, Ramachandrapuram, India, January 21, 2009.

140 *if you put air-conditioning in your office:* Gary Mormino, interview by author, St. Petersburg, FL, December 8, 2008.

140 *Kusam Raja Mouli:* Kusam Raja Mouli, interview by author, January 2009.

140 *The McKinsey Global Institute:* Ablett et al., "The 'Bird of Gold.'"

141 *makers of designer luxuries like $3,000 handbags:* Anand Giridharadas, "Western-Style Luxe May Be a Hard Sell for Modern India," *New York Times*, March 25, 2009.

141 *masala dosas:* A popular South Indian food, the masala dosa is a thin pancake made of rice and legume flour wrapped around a spiced filling of potatoes, chilies, and onions.

141 *found that 40 percent of the metropolitan area's electricity:* Chittaranjan

Tembhekar, "ACs eat up 40% of city's total power consumption," *Times of India,* December 23, 2009.

141 *Girish Srinivasan:* Girish Srinivasan, interview by author, Mumbai, December 16, 2008.

142 *1 to 2 percent of Indian households:* Keith Bradsher, "The Price of Keeping Cool in Asia; Use of Air-Conditioning Refrigerant Is Widening the Hole in the Ozone Layer," *New York Times,* February 23, 2007.

143 *RUPE estimates:* Research Unit for Political Economy, "India's Runaway Growth: Distortion, Disarticulation, and Exclusion," *Aspects of India's Economy,* April 2008.

144 *The vast majority of the workforce:* Ibid.

144 *private cars meet less than 5 percent:* Ibid.

144 *six air-conditioned shopping malls:* Ablett et al., "The 'Bird of Gold.' "

146 *Keith Dias:* interview by author, Hyderabad, January 15, 2009.

147 *S. Srinivas:* interview by author, Hyderabad, January 23, 2009.

149 *Hariharan Chandrashekar:* interview by author, Hyderabad, January 21, 2009.

Chapter 8

150 *You've got your ice:* Paul Theroux, *The Mosquito Coast* (Boston: Houghton Mifflin, 1982), 38–39.

150 *The chief source of problems:* Eric Sevareid, CBS News, December 29, 1970, quoted in Thomas Martin, *Malice in Blunderland* (New York: McGraw-Hill, 1973), 23.

150 *Nicholas Georgescu-Roegen:* Nicholas Georgescu-Roegen, *The Entropy Law and the Economic Process* (Cambridge, MA: Harvard University Press, 1971), 276–91.

151 *Capitalist economies:* One of the clearest explanations of the consequences of the growth imperative is in John Bellamy Foster and Brett Clark, "The Paradox of Wealth: Capitalism and Ecological Destruction," *Monthly Review* 61, no. 6 (2009), 1–18.

151 *Why do things get weaker and worse?:* Theroux, *The Mosquito Coast,* 155, 194.

152 *a 2008 analysis by Minqi Li:* Minqui Li, "Climate Change, Limits to Growth, and the Imperative for Socialism," *Monthly Review,* 60, no. 6 (2009), 51–67.

152 *The U.S. Department of Energy reported:* Energy Information Agency, "U.S. Energy-Related Carbon Dioxide Emissions Declined by 2.8 Percent in 2008," 20 May 2009, http://www.eia.doe.gov/neic/press/press318.html

#2009_05_20; Timothy Garnder, "U.S. CO_2 Emissions from Fuels to Fall 5 Pct in 2009," Reuters, August 11, 2009; EIA, "Annual Energy Review 2008," 2008, http://www.eia.doe.gov/emeu/aer/envir.html.

153 *we could produce a 100-pound cake:* Herman Daly, *Ecological Economics and Sustainable Development: Selected Essays by Herman Daly* (Cheltenham, UK: Edward Elgar Publishers, 2008), 44.

153 *Both the theory and the experience:* Foster and Clark, "The Paradox of Wealth."

153 *The EPA estimated in 2006:* U.S. Environmental Protection Agency, "Report to Congress on Server and Data Center Energy Efficiency: Public Law 109–431," August 2, 2007, www.energystar.gov/ia/partners/prod_development/downloads/EPA_Datacenter_Report_Congress_Final1.pdf.

154 *The lion's share:* Neil Rasmussen, "Calculating Total Cooling Requirements for Data Centers," American Power Conversion White Paper no. 25, 2003, http://www.apcmedia.com/salestools/NRAN-5TE6HE_R2_EN.pdf.

154 *can consume as much power as would thirty typical California homes:* EPA, "Report to Congress."

154 *In line with 1960s-era predictions:* Kenneth Brill, "The Invisible Crisis in the Data Center: The Economic Meltdown of Moore's Law," Uptime Institute White Paper, 2007, http://uptimeinstitute.org/wp_pdf/(TUI3008)Moore'sLawWP_080107.pdf.

154 *The industry continues to invest:* Wataru Nakayama, "Exploring the Limits of Air Cooling," *ElectronicsCooling* 12, no. 3 (2006), 10–17; Roger Schmidt, "Challenges in Electronic Cooling—Opportunities for Enhanced Thermal Management Techniques—Microprocessor Liquid Cooled Minichannel Heat Sink," *Heat Transfer Engineering,* 25 (2004), 3–12.

155 *paper consumption:* World Resources Institute, "Resource Consumption: Paper and Paperboard Consumption Per Capita," 2007, http://earthtrends.wri.org/searchable_db/index.php?theme=9.

155 *U.S. business travel marched upward:* Christophe Renard, "Business Travel Drives Economic Growth," *CWT Vision,* September 2007.

155 *revenues from Internet advertising almost quadrupled:* Pricewaterhouse Coopers LLP, 'IAB Internet Advertising Revenue Report,' March 2009, http://www.iab.net/media/file/IAB_PwC_2008_full_year.pdf

156 *25 percent of all traffic through search engines goes to retail sites:* "Retail Sites Get 25% of Traffic from Search Engines," Marketing Charts, June 2008, http://www.marketingcharts.com/interactive/retail-sites-get-25-of-traffic-from-search-engines-588.

156 *Residential electricity use has increased by 23 percent:* EIA, "Annual Energy Outlook 2009."

157 *Looking at the evolution of cars in time:* John M. Polimeni, Kozo Mayumi, Mario Giampietro, and Blake Alcott, *The Jevons Paradox and the Myth of Resource Efficiency Improvements* (London: Earthscan, 2008), 79–140.

157 *If an explicit political decision is not made:* For my angle on this issue, see Stan Cox, *Sick Planet: Corporate Food and Medicine* (London: Pluto Press, 2008), 154–75.

157 *to increase efficiency now:* Polimeni et al., *Jevons Paradox,* 80.

158 *predicament of efficiency:* Mithra Moezzi, "The Predicament of Efficiency," Proceedings of the 1998 ACEEE Summer Study on Energy Efficiency in Buildings, Washington, DC, August 1998, 4.273–82, http://enduse.lbl .gov/info/ACEEE-Pred.pdf.

158 *two refrigerators:* Don Hopey, "Energy Stars May Not Be All They Say They Are," *Pittsburgh Post-Gazette,* November 9, 2008. I compared the models reported on by Hopey with Energy Star ratings of other models, found at http://www.energystar.gov/index.cfm?c=refrig.pr_crit_refrig erators.

158 *such as a twenty-two-cubic-foot, freezer-on-top Maytag:* EPA/DOE, "Find Energy Star Qualified Refrigerators and Freezers," 2008, http://www .energystar.gov/index.cfm?fuseaction=refrig.search_refrigerators.

159 *One of the difficulties in moralizing:* Moezzi, "Predicament of Efficiency."

159 *In concentrating on* efficiency: Elizabeth Shove, "Efficiency and Consumption: Technology and Practice," *Energy and Environment* 15 (2004), 1053–65.

160 *an impressive 28 percent more efficient:* I estimated average SEER from figures in Stephen Meyers, James McMahon, and Michael McNeil, "Realized and Prospective Impacts of U.S. Energy Efficiency Standards for Residential Appliances: 2004 Update," June 24, 2005, Lawrence Berkeley National Laboratory Report No. LBNL-56417.

160 *Jevons in his book* The Coal Question: William Stanley Jevons, *The Coal Question: An Inquiry Concerning the Progress of the Nation, and the Probable Exhaustion of Our Coal Mines,* 2nd ed. (London: Macmillan and Co., 1866). Digital copy retrieved from http://www.eoearth.org/article/The _Coal_Question_%28e-book%29.

160 *rebound and backfire:* Blake Alcott, "Jevons' Paradox," *Ecological Economics* 54 (2005), 9–21; Polimeni et al., *Jevons Paradox.* In the latter, Alcott writes, "Today's environmental efficiency strategy claims that an input's more efficient use lowers its rate of consumption. The inverse/corollary

of this is that were processes to become *less* efficient, we would consume the input at a *higher* rate. Or had technological efficiency increase remained unchanged—stopped, say, at around 1781 [with the Watt steam engine] we would, according to the strategy's assumptions, today consume a hundred or a thousand times as much—or infinitely more—labor or cotton or fuel than we do today after over two centuries of efficiency increase. To maintain that rebound is less than 100 percent one must defend this conclusion. . . . If we take time, material, energy and space inputs and assume all historically known efficiencies away, we most likely arrive at the population and per capita production of hunter-gatherer societies living sustainably" (47–48).

161 *We saw an example of apparent backfire:* Jevons's paradox was at work even four hundred years before Jevons. Todd Bostwick says the Hohokam people of central Arizona "became so efficient in building and operating canals, they got carried away and greatly overdesigned their canal system. That contributed to their collapse."

161 *Government programs for insulating:* Lance McCold, Richard Goeltz, Mark Ternes, and Linda Berry, "Texas Field Experiment: Performance of the Weatherization Assistance Program in Hot-Climate, Low-Income Homes," Oak Ridge National Laboratory report ORNL/CON-499 (April 2008), http://weatherization.ornl.gov/pdf/CON499.pdf

161 *engineering models often assume that a given percentage:* Jeffrey Dubin, Allen Miedema, and Ram Chandran, "Price Effects of Energy-Efficient Technologies: A Study of Residential Demand for Heating and Cooling," *RAND Journal of Economics* 17 (1986), 310–25.

162 *estimates of 0 to 50 percent for air-conditioning:* Lorna Greening, David Greene, and Carmen Difiglio, "Energy Efficiency and Consumption—The Rebound Effect—A Survey," *Energy Policy* 28 (2000), 389–401.

162 *how do people respond:* See, for example, JEA, formerly Jacksonville (FL) Electric Authority, "Compact Fluorescent Light Bulbs (CFL)," 2009, http://www.jea.com/community/conservcenter/home/cfl.asp; Energy Star, "Compact Fluorescent Light Bulbs," 2009, http://www.energystar.gov/index.cfm?c=cfls.pr_cfls.

162 *efficiency gains and rates of resource use:* For some whole-economy examples of efficiency and consumption rising hand-in-hand, see Polimeni et al., *Jevons Paradox*, 141–71.

163 *mainstream economists have often denied the existence of the paradox:* Prominent among critics of rebound/backfire analyses has been efficiency prophet Amory Lovins of the Rocky Mountain Institute in Snowmass, Colorado. Lovins's argument was well eviscerated by Robert Bryce.

See Amory Lovins, "Energy Saving from More Efficient Appliances: Another View," *Energy Journal* 9 (1988), 155–62; Robert Bryce, "Green Energy Advocate Amory Lovins: Guru or Fakir?" *Energy Tribune*, November 12, 2007, http://www.energytribune.com/articles.cfm?aid=676. Look for Bryce's striking graph comparing the steep decrease in energy use per dollar of gross domestic product with the steep increase in energy consumption.

163 *Maximizing energy efficiency has no particular merit:* Leonard Brookes, "Energy Efficiency Fallacies Revisited," *Energy Policy* 28 (2000), 355–66.

164 *DST:* Matthew Kotchen and Laura Grant, "Does Daylight Saving Time Save Energy? Evidence From a Natural Experiment in Indiana," Working Paper No. 14429, National Bureau of Economic Research, 2008, http://www.nber.org/papers/w14429.

164 *even some environmental organizations and environmentalists:* See, for example, Environmental Defense Fund, "Questions and Answers on Nuclear Power," September 9, 2008, http://www.edf.org/article.cfm?contentid=4470, and Alexis Madrigal, "Co-Founder of Greenpeace Envisions a Nuclear Future," *Wired*, November 19, 2007.

165 *all 103 existing nuclear reactors:* 2007 nuclear generation data are from Energy Information Agency, "U.S. Nuclear Generation of Electricity," 2009, http://www.eia.doe.gov/cneaf/nuclear/page/nuc_generation/gensum.html.

165 *In 2008, Joshua Pearce:* Joshua Pearce, "Thermodynamic Limitations to Nuclear Energy Deployment as a Greenhouse Gas Mitigation Technology," *International Journal of Nuclear Governance, Economy and Ecology* 2 (2008), 113–30. Pearce also shows that much of nuclear power's own usage is spent getting fissionable uranium isotopes out of uranium ore, which is dilute, containing anywhere from a fraction of a percent to a few percent uranium oxide (U_3O_8). Furthermore, it is unclear how much of the planet's useful uranium ore, and how much of the radioactive isotope ^{235}U, remains to be mined. Without further growth, currently known reserves would keep the global nuclear industry running easily beyond 2050, however. But with strong growth, says the International Atomic Energy Agency, nuclear generation could exhaust today's known, affordable uranium reserves before 2050, with deficits arising as early as 2026. "Very high-cost" reserves and "unconventional" reserves—both costly in energy as well as money terms—would be needed to keep the industry growing. New, high-quality uranium deposits surely remain to be discovered, but it has been more than twenty years since "world class" deposits containing large quantities of high-quality ore have been discovered. As

might be expected, the bigger the deposit, the lower the quality, on average; any significant new deposits will probably lie deeper underground than current ones and be more expensive to mine.

166 *Breeder reactors:* Joshua Pearce, author interview by telephone, February 13, 2009.

167 *a fire resulting from an accident:* Robert Alvarez et al., "Reducing the Hazards from Stored Spent Power-Reactor Fuel in the United States," *Science and Global Security* 11 (2003), 1–51.

167 *vitrified before storage:* Vitrification won't work with plutonium; the element has to be removed from wastes that are to be vitrified. Until recently, it was hoped that a ceramic material called zircon (a compound containing zirconium, silicon, and oxygen) could be laced with 10 percent plutonium and remain stable for tens of thousands of years. But recent research found that radiation from the plutonium would degrade zircon's structure in only 1,400 years: Ian Farnan, Herman Cho, and William Weber, "Quantification of Actinide μ-Radiation Damage in Minerals and Ceramics," *Nature* 445 (2007), 190–93.

168 *reprocessing:* Solveig Torvik, "The French Fix," *Seattle Post-Intelligencer,* April 22, 1998; Shaun Burnie, "French Nuclear Reprocessing—Failure at Home, Coup D'etat in the United States," Public Citizen, May 2007, http://www.citizen.org/cmep/energy_enviro_nuclear/nuclear_power_plants.

169 *the government of France:* Julio Godoy, "European Heat Wave Shows Limits of Nuclear Energy," CommonDreams.org, July 28, 2006, http://www.commondreams.org/headlines06/0728-06.htm.

169 *American Electric Power Co. shut down:* Shannon Harrington, "U.S. Heat Wave Heads to Northeast, May Break Records," Bloomberg News, July 31, 2006.

169 *the Tennessee Valley Authority shut down:* Beth Rucker, "Heat Wave Kills 41 in South, Midwest," Associated Press, August 17, 2007.

169 *warming of the atmosphere:* This effect has been seen in rivers and lakes in Europe. See R. E. Hari, D. M. Livingstone, R. Siber, P. Burkhardt-Holm, and H. Güttinger, "Consequences of climatic change for water temperature and brown trout populations in Alpine rivers and streams," *Global Change Biology* 12 (2006), 10–26; D. M. Livingstone, "Impact of secular climate change on the thermal structure of a large temperate central European lake," *Climatic Change* 57 (2003), 205–225.

170 *9 billion gallons of ethanol fuel:* Because it contains less energy per volume, that quantity of ethanol has the same energy content as 6 billion gallons

of gasoline. At least 7 billion gallons are currently used in vehicle air-conditioning (see chapter 3).

170 *enough fuel ethanol to substitute for about 70 percent:* three hundred fifty million acres times 155 bushels per acre times a generally reported 2.8 gallons of ethanol per bushel equals 152 billion gallons of ethanol. With ethanol's lower energy content, that is equivalent to 100 billion gallons of gasoline. The United States consumed 142 billion gallons of gasoline in 2007.

170 *destruction of irreplaceable soil:* See Jerry Glover, Cindy Cox, and John Reganold, "Future Farming: A Return to Roots?" *Scientific American,* August 2007.

170 *A U.S. Department of Energy/Department of Agriculture blueprint:* R.D. Perlack et al., *Biomass as a Feedstock for a Bioenergy and Bioproducts Industry: The Technical Feasibility of a Billion-Ton Annual Supply* (Oak Ridge, TN: Oak Ridge National Laboratory, 2005), available at feedstockreview.ornl.gov/pdf/billion_ton_vision.pdf; also see Tadeusz Patzek, "The Cellulosic Ethanol Delusion," *Energy Tribune,* June 14, 2007, http://www.energytribune.com/articles.cfm?aid=516.

Chapter 9

172 *The basic tenet of the adaptive model:* Richard de Dear and Gail Brager, "The Adaptive Model of Thermal Comfort and Energy Conservation in the Built Environment," *International Journal of Biometeorology* 45 (2001), 100–108.

172 *occupy entire books:* See, for example, Mat Santamouris, ed., *Advances in Passive Cooling* (London: Earthscan, 2007).

172 *green energy bubble:* See Eric Janszen, "The Next Bubble," *Harper's,* February 2008.

173 *energy savings of up to 25 percent:* Henry Nasution and Mat Nawi Wan Hassan, "Potential Electricity Savings by Variable Speed Control of Compressor for Air-conditioning Systems," *Clean Technology and Environmental Policy* 8 (2006), 105–11.

173 *Variable-speed electric motors:* Haruyuki Fujii and Loren Lutzenhiser, "Japanese Residential Air-Conditioning: Natural Cooling and Intelligent Systems," *Energy and Buildings* 18 (1992), 221–33.

174 *Surveys of apartment residents:* Loren Lutzenhiser, "A Question of Control: Alternative Patterns of Room Air-Conditioner Use," *Energy and Buildings* 18 (1992), 193–200; Willett Kempton, Daniel Feuermann, and

Arthur McGarity, " 'I Always Turn It on Super': User Decisions about When and How to Operate Room Air Conditioners," *Energy and Buildings* 18 (1992), 177–91.

175 *ducted air-conditioning systems are generally not desired:* Fujii and Lutzenhiser, "Japanese Residential Air-Conditioning."

175 *intermittent cooling:* Sometimes, says Raymond Dean, the opposite strategy works better, however: "When crowds will exist for only short periods, as in a church, and when there is substantial thermal mass, it's common practice to precool the space for a long time beforehand, using a low-capacity system. Then the building's large thermal mass can absorb most of the heat generated while the space is occupied." Interview by author, August 9, 2009.

177 *Typical savings:* Progress Energy, Raleigh, NC, "Water Heater," 2009, http://www.progress-energy.com/custservice/carres/energytips/hotwater .asp.

177 *paperless offices:* But the additional electronic equipment in a paperless office could generate enough heat to cancel out any energy savings from using fan-assisted cooling.

178 *energy savings of 15 to 40 percent:* Hashem Akbari, Ronnen Levinson, William Miller, and Paul Berdahl, "Cool Colored Roofs to Save Energy and Improve Air Quality," Lawrence Berkeley National Laboratory Paper LBNL-58265, 2005.

178 *Akbari skirted:* H. Akbari, M. Pomerantz, and H. Taha, "Cool Surfaces and Shade Trees to Reduce Energy Use and Improve Air Quality in Urban Areas," *Solar Energy* 70 (2001), 295–310.

179 *For maximum conservation:* James Simpson and Gregory McPherson, "Potential of Tree Shade for Reducing Residential Energy Use in California," *Journal of Arboriculture* 22 (1996), 10–18.

179 *Covering a roof:* T. Takakura, S. Kitade, and E. Goto, "Cooling Effect of Greenery Cover over a Building," *Energy and Buildings* 31 (2000), 1–6.

180 *direct load control:* Willett Kempton, Cathy Reynolds, Margaret Fels, and David Hull, "Utility Control of Residential Cooling: Resident-Perceived Effects and Potential Program Improvements," *Energy and Buildings* 18 (1992), 201–19.

180 *This utility ploy:* Raymond Dean, interview by author, August 9, 2009.

180 *Yolande Strengers:* Strengers, "Comfort Expectations."

181 *the average household spends:* Spending figures are from EIA, "Households, Buildings, Industry and Vehicles," 2009, http://www.eia.doe.gov /emeu/consumption/index.html.

181 *Whenever I suggest:* Raymond Dean, interview by author, August 9, 2009.

182 *Although frustrated:* Strengers, "Comfort Expectations."

183 *ideally, we would democratize access:* Eric Klinenberg, testimony before the California Senate Governmental Organization Committee, August 9, 2006.

183 *A few local governments:* Lisa Gray, "Limits on Living Large," *Houston Chronicle*, May 30, 2008.

184 *cuts in our dependence on private vehicles:* Ray Dean brings up one unexpected potential consequence of energy-saving strategies: "Might constraints placed on allowed temperature levels in homes, office buildings, schools, and stores, motivate people to drive around more in airconditioned cars, just to get cool? Although it might sound crazy, it's worth asking: How does the environmental cost of air-conditioning one or two people in a single automobile actually compare with the environmental cost of air-conditioning one or two people in their centrally-cooled home? Although the fuel used may be worse, the mechanical efficiency may be worse, and the insulation may be worse, the total surface area of an automobile is so much smaller than the total surface area of a typical residence that I wouldn't be surprised if automobile air-conditioning is more benign than conventional residential air-conditioning for cooling the same number of people." Interview by author, June 2009.

184 *cooling their seats:* Herb Shuldiner, "Air-Conditioned Seats Can Save on Fuel Economy," *Sun-Journal* (Lewiston, Maine), February 17, 2007.

184 *The EPA, with an international consortium:* EPA, "Mobile Air-Conditioning," 2009, http://www.epa.gov/cppd/mac/.

185 *The promotion of technical efficiency:* Elizabeth Shove, "Efficiency and Consumption."

185 *heat wheels:* Office of Energy Efficiency and Renewable Energy, U.S. Department of Energy, "Desiccant Dehumidification," publication GO-102001-1165, May 2001; Ray Dean, interview by author, August 2009.

186 *solar devices:* Hans-Martin Henning, "Solar Assisted Air-conditioning of Buildings—An Overview," *Applied Thermal Engineering* 27 (2007), 1734–49.

186 *Even Albert Einstein:* A. Einstein et al., "Refrigerator," U.S. Patent 1,781,541, November 11, 1930, copy at http://www.fourmilab.ch/etexts/einstein/uspat1781541/www.

186 *Detailed descriptions of such systems are complex:* Einstein and Szilárd's design used butane, ammonia, and water. In most current absorption cycles for air-conditioning, a solution of water and lithium bromide is heated, boiling off the water, which enters a condensing chamber where

it is cooled and liquefied. The water then moves into an evaporator chamber under low pressure, where it can absorb heat from other water that's passing through pipes. The piped water will go to cool a space, while the water that took on its heat boils again under the low pressure. As steam, it goes into a fourth chamber, where it is reabsorbed into a strong lithium bromide solution, which is returned to the start of the cycle.

186 *Sorption systems are especially well suited:* A good review is A.M. Papado-poulos, S. Oxizidis, and N. Kyriakis, "Perspectives of Solar Cooling in View of the Developments in the Air-Conditioning Sector," *Renewable and Sustainable Energy Reviews* 7 (2003), 419–38.

186 *a photovoltaic-powered conventional air conditioner:* The efficiency of an absorption system is the efficiency of the solar collector efficiency (about 50 percent) times that of the absorption apparatus (about 60 percent), which comes to around 30 percent total efficiency. For a solar photovoltaic-powered mechanical unit, the photovoltaic panel efficiency (about 15 percent) times that of a high-efficiency conventional air-conditioning unit (about 400 percent) comes to about 60 percent. Information from Raymond Dean, interview by author, August 9, 2009.

186 *The largest facility in American currently being cooled:* Bob Livingston, in-terviewed by author, Phoenix, AZ, July 17, 2009.

187 *Researchers in Florida:* B.S. Davanagere, S.A. Sherif, and D.Y. Goswami, "A Feasibility Study of a Solar Desiccant Air-Conditioning System—Part I: Psychrometrics and Analysis of the Conditioned Zone," *International Journal of Energy Research* 23 (1999), 7–21.

188 *Ground-source heat pumps:* J. Lund, B. Sanner, L. Rybach, R. Curtis, and G. Hellstrom, "Geothermal (Ground-Source) Heat Pumps: A World Overview," *Geo-Heat Center Bulletin* (Oregon Institute of Technology) 25, no. 3 (2004), 1–10.

188 *Biodiversity Conservation India Limited:* Anuradha Desikan Eswar, BCIL, Bangalore, India, e-mail interview by author, January 25, 2009.

189 *A round tower made largely of fabric:* E. Erell, D. Pearlmutter, and Y.A. Etzion, "A Multi-Stage Down-Draft Evaporative Cool Tower for Semi-Enclosed Spaces: Aerodynamic Performance," *Solar Energy* 82 (2008), 420–29.

189 *Torrent Research Centre:* L.E. Thomas and G. Baird, "Post-Occupancy Evaluation of Passive Downdraft Evaporative Cooling and Air-Conditioned Buildings at Torrent Research Centre, Ahmedabad, India," in *Challenges for Architectural Science in Changing Climates—Proceedings of the 40th Annual Conference of the Architectural Science Association*

ANZAScA, ed. S. Shannon, V. Soebarto, and T. Williamson (Adelaide: University of Adelaide and The Architectural Science Association ANZAScA, 2006), 97–104.

190 *solar chimney:* N.K. Bansal, R. Mathur, and M.S. Bhandari, "Solar Chimney for Enhanced Stack Ventilation," *Building and Environment* 28 (1993), 373–77.

190 *hybrid ventilation systems:* Reviewed by P. Wouters, N. Heijmans, C. Delmotte, and L. Vandaele, "Classification of Hybrid Ventilation Concepts," Belgian Building Research Institute, August 26, 1999, http://hybvent.civil .auc.dk/puplications/report/19990826_classification_Hybvent_concepts _pw.PDF; de Dear and Brager, "The Adaptive Model."

192 *bought a new stove:* Herman Daly, *Steady State Economics,* 2nd ed. (Washington, DC: Island Press, 1991), 119.

192 *Reasons for the popularity:* Helmut Feustel, Anibal de Almeida, and Carl Blumstein, "Alternatives to Compressor Cooling In Residences," *Energy and Buildings* 18 (1992), 269–286.

193 *people who live or work in naturally ventilated buildings:* Richard de Dear and Gail Brager, "Thermal Comfort in Naturally Ventilated Buildings: Revisions to ASHRAE Standard 55," *Energy and Buildings* 34 (2002), 549–61.

193 *building occupants are not simply:* de Dear and Brager, "Adaptive Model."

194 *Standard 55 was expanded:* See ASHRAE, *ASHRAE Standard 55-2004— Thermal Environmental Conditions for Human Occupancy* (Atlanta: ASHRAE, 2004).

194 *Figure 4:* de Dear and Brager, " "The Adaptive Model."

195 *Fergus Nicol:* Fergus Nicol, "Adaptive thermal comfort standards in the hot-humid tropics," *Energy and Buildings* 36 (2004), 628–637.

195 *building occupants, engineers, and designers:* Heather Chappells and Elizabeth Shove, "Debating the Future of Comfort: Environmental Sustainability, Energy Consumption and the Indoor Environment," *Building Research and Information* 33 (2005), 32–40.

196 *an innovative and inventive trajectory:* Stephen Healy, "Air-Conditioning and the 'Homogenization' of People and Built Environments," *Building Research and Information* 36 (2008), 312–22.

196 *thermal sense:* Lisa Heschong, *Thermal Delight in Architecture* (Cambridge, MA: MIT Press, 1979).

INDEX